FOREWORD

A companion web site has been established to go along with this report so that readers of this report may view pictures, maps and graphs in color and link direct to referenced web sites.

The URL to this "Master" link is:

http://natgas-rpt.com

This book is being compiled under the auspices of Western Research Institute, Inc. (WRI) WRI will continue to research and compile facts about natural gas after this report is published. These facts will be published on the above companion web site under the headings "Updates" and "Supplements."

A distinction is made in this report between "imported oil" and "imported foreign oil." "Imported oil" is oil imported from Canada and Mexico.

"Imported foreign oil" is oil imported from outside of North America, mainly OPEC countries

THIS REPORT SEEKS TO ANSWER:

1. Whether and to what degree there are forces and entities that are maintaining the price of gasoline and diesel at artificially high levels?

2. Whether and how fast natural gas production can replace imported foreign oil?

3. How fast can conversion to natural gas as a fuel for transportation be accomplished?

4. What are the obstacles to conversion to natural gas as a fuel for transportation?

5. Are there people and entities keeping it from happening?

6. What are the benefits of converting to natural gas as a fuel for transportation?

7. How much refining is required for natural gas?

8. What are the components of raw natural gas and their value on the market place for transportation and manufacturing purposes?

9. What states are benefiting from production of natural gas?

10. How can all states benefit from conversion to natural gas?

An Investigative Report

HOW AMERICA CAN STOP IMPORTING FOREIGN OIL & THOSE PREVENTING IT FROM HAPPENING

Jerry Fenning & Charles Hoppins

Western Research Institute, Inc.

Print Edition ISBN 978-1-882567-53-9

987654

PREFACE

Much of this Report is statistical and consists of facts from many different sources. All the facts reported are referenced as to source. Facts compiled have mostly come from Internet web sites and will continue to be compiled even after this report is published.

There have been many challenges in the compiling of this report. One challenge is verifying the accuracy of statistics that are reported. There were occasional conflicting statistics reported even on the same web site.

Another has been the frequency of changes in the web pages that are referenced, which renders the links to the referenced URLs inoperable. In addition statistics are often updated. Thus the statistics reported in this published report may not match those referenced. Fortunately this will be compensated for with the maintaining of a companion web site, which has an end notes page which can be updated to match current statistics. This may result in a disparity between end notes in the published report and end notes in the companion web site.

There are many government and non-government web sites that provide statistics. The main one is the Energy Information Administration (EIA), which administers a massive database on energy related matters. It is a division of the U.S. Department of Energy.[1]

Other web sites that should be of interest include the Environmental Protection Agency (EPA), the Government Printing Office (GPO) and the Library of Congress (LOC).[2-4]

Exports of oil from the U.S. are increasing every year. More than one billion barrels were exported from the U.S. in 2010, according to the EIA. EIA Tables illustrate how imports are declining, how oil production in the U.S. is increasing, how exports are increasing, how consumption of petroleum is decreasing, and how production and consumption of natural gas is increasing.

END NOTES:

A companion web site has been established to go along with this report so that readers may view pictures, graphs and maps in color and link direct to referenced web sites.

The URL to this "Master" link is:

http://natgas-rpt.com

1 U. S. Energy Information Agency (2012, August). Retrieved from **http://www.eia.gov/**

2 U. S. Environmental Agency (EPA), (2012, August). Retrieved from **http://www.epa.gov**

3 U. S. Government Pringint Office, Keeping America Informed, (2012, August). Retrieved from **http://www.gpo.gov**

4 Library of Congress, (2012, August). Retrieved from **http://www.loc.gov**

TABLE OF CONTENTS

Foreword   3
Title Page   5
Copyright Page   6
Preface   7
Table of Contents   9
Executive Summary   11

CHAPTER

1. INTRODUCTION   13

II AMERICA'S PROBLEMS   23

III. TERRORISM AND WAR   27

IV. POLLUTION AND GLOBAL WARMING   33

V. TRANSFER OF WEALTH   41

VI. NATURAL GAS DISCOVERIES   49

VII. PROS AND CONS OF FRACTURING   57

VIII. CONVERTING TO NATURAL GAS   67

IX. NATURAL GAS HIGHWAY   83

X. FAILING GRADE FOR CONVERSION   89

XI. PUBLIC/PRIVATE PARTNERSHIP   97

XII. NATURAL GAS ACT   103

XIII WHAT IF   109

XIV LIGHT DUTY VEHICLES   115

APPENDIXES

A. Epilogue   119
B. Comparative Factors   121
C. Acknowledgments   127
D Definitions   129
E Natural Gas Act   133
F. Western Research   179
G. About The Authors   181

EXECUTIVE SUMMARY
(Findings of Fact and Conclusions of Logic)

There are items that have stood out in compiling this report. They are:

- The severe consequences that resulted from the exemption in 2005 of the oil and gas companies from complying with the Clean Water Act under the Bush/Cheney Administration.

- The massive drilling campaign within the continental U.S. resulting from discovery of new shale oil fields or plays and the development of horizontal drilling techniques and fracturing resulting in an oversupply of natural gas.

- The huge potential benefits of implementing natural gas as an alternative fuel for transportation.

- The degree of corruption within the oil and gas industry and the Congress.

Conclusions by Western Research Institute include:

- Increased production of crude oil within the continental U.S. and Alaska will have little or no effect on crude oil prices, gasoline or diesel prices or U.S. energy independence.

- Converting to natural gas for transportation can make America energy independent (if done on a massive scale)

- The U.S. can be totally free of foreign oil imported from outside of North America in five to ten years and keep trillions of dollars in circulation in the U.S.

- The U.S. can have hundreds of thousands of natural gas vehicles on the road in five years, a million natural gas vehicles on the road in ten years and millions on the road in succeeding years.

- Converting to natural gas for transportation can create millions of jobs and benefit the environment (if done on a massive scale).

- But it won't happen because the petroleum industry and the U.S. Congress won't let it happen, unless the American people make it happen. And the main reason the American people will not make it happen is lack of public awareness

CHAPTER I
INTRODUCTION

The world is becoming more energy intensive as the industrial and computer revolutions evolve and as standards of living increase in many countries. Using natural gas as a transportation fuel is a substantial departure from an oil-driven economy. The world markets for oil have become behemoth in size and exceedingly complex in scope. Supplies of natural gas in the U.S. are currently (summer of 2012) exceeding demand and are offering an attractive alternative as a fuel for transportation.

The price of crude oil and petroleum products are determined by a number of factors. These factors include supply and demand, competition, trading in commodity markets and government policies.

The commodity markets are where futures contracts for crude oil and petroleum products are bought and sold. Supply and demand supposedly determines the price of crude and other petroleum products. The price of crude oil is related to the price Americans pay for gasoline and diesel.

Supplies of crude and other petroleum products are manipulated through variations in production. The price of crude and other petroleum products are also manipulated on the world's commodity exchanges.[1]

There are many commodity exchanges around the world. The CME Group Inc headquartered in Chicago, the world's largest commodity exchange, owns and operates commodity exchanges in Chicago and New York. They are Chicago Mercantile Exchange (CME), Chicago Board of Trade (CBOT), the New York Mercantile Exchange (NYMEX), and the Commodity Exchange (COMEX).

The NYMEX is where energy futures are traded. These include 11 classes of Crude, 10 classes of natural gas, and 10 classes of refined petroleum products such as gasoline, heating oil and diesel fuel. Other energy futures traded on the NYMEX are electricity, ethanol, coal and uranium.[2]

REGULATING COMMODITY TRADING

The U.S. government agency that regulates commodity trading is the Commodities Futures Trading Commission. "The CFTC's mission is to protect market users and the public from fraud, manipulation, abusive practices and systemic risk related to derivatives that are subject to the Commodity Exchange Act, and to foster open, competitive, and financially sound markets."[3]

Although the supply and price of oil are manipulated via commodity exchanges, the primary factor that determines the price of petroleum products still rests on the oil companies. Oil is produced in many countries throughout the world. For a time in 2012, the largest private company in the United States and the largest non-government owned oil company was ExxonMobile. It was only ranked 17th among the largest oil companies in the world because the other 16 larger companies are all national entities which are owned by their respective governments.

It is important to recognize that the policies of these nationalized oil companies often are instruments of their government. The largest oil companies are ranked below in order of size of total reserves of barrels of oil and natural gas equivalent as of 2011. The rankings are subject to change. Private companies are italicized.[4] They are:

1. Saudi Arabian Oil Co.,   303 billion
2. National Iranian Oil Co.,   300 billion
3. Qatar General Petroleum Corp.   170 billion
4. Iraq National Oil Co.,   134 billion
5. Petroleos de Venezuela, S.A.,   128 billion
6. Abu Dhabi National Oil Co. (UAE),   126 billion
7. Kuwait Petroleum Corp.,   111 billion
8. Nigerian National Petroleum Corp.,   68 billion
9. National Oil Co. (Libya),   50 billion
10. Sonatrach (Algeria),   39 billion
11. Gazprom (Russia),   29 billion
12. OAO Rosneft (Russia),   21 billion
13. PetroChina Co. Ltd.,   21 billion
14. Petronas Malaysia,   19 billion
15. OAO Lukoil (Russia),   16 billion
16. Egyptian General Petroleum Corp.,   14 billion
17. *ExxonMobil Corp.,*   *13 billion*
18. *Petroleos Mexicanos,*   13 billion
19. *BP Corp. (UK),*   *12.5 billion*
20. Petroleo Brasilerio S.A. (Brazil),   11.5 billion
21. *Chevron Corp. (U.S.),*   *10.8 billion*
22. *Royal Dutch Shell,*   *10.7 billion*
23. *ConocoPhillips (U.S),*   *10.6 billion*

All the first 10 oil companies listed are members of OPEC which is a major force in controlling the price of crude oil, mostly by limiting supply. Changes in supply are achieved by varying the amount of oil that is produced or pumped from their fields. Another strategy is to withhold oil from entering the open market by storing it until more favorable conditions arise.

ExxonMobil, BP, Chevron Corp., Royal Dutch Shell and ConcocoPhillips comprise the five "supermajor" oil companies that control much of the market in the United States. It is these five that comprise what is often referred to as "Big

Oil." BP has been caught trying to manipulate the commodities market on the NYMEX. Trading futures makes up a significant percentage of BP's revenue.[5]

In his book and on his Blog on the Huffington Post, Raymond Learsy says:

> Permit me to quote once again the words of Leon Hess, founder of Hess Oil... He spoke before a Senate Committee on Government Affairs some 20 years back. His words were as true then as they are now, even incorporating all the trading exchanges that have blossomed around the world: "I am an old man, but I would bet my life that if the Merc (the Nymex) was not in operation there would be ample oil and reasonable prices all over the world without this volatility."
>
> ...Clearly the commodity exchanges are subject to being manipulated and have been, and in all likelihood are continuing to be manipulated. Consider that more than 137 billion barrels of oil were traded on the NYMEX alone last year (2009)...Yet the world consumes barely 30 billion barrels of oil annually.
>
> How many Wall Street or London or Singapore bank oil trading desks with no interest in consuming or producing oil, but with wide access to banking resources and to oil company trading intelligence are going along for the profitable ride.
>
> ... And who pays the bill? Yes, you guessed it, you do. Not only at vast economic cost, but at grave risk to our national security.
>
> Thanks for the lesson BP![6]

There are a number of large trading houses around the world that "amass speculative positions worth billions in raw goods, or hoard commodities in warehouses and super tankers during periods of tight supply."[7]

Among those involved in trading oil are Vitol Group of Geneva and Rotterdam; Koch Industries, Inc., headquartered in Wichita, Kansas; Trafigura, headquarted in Amsterdam; Mercuria, headquartered in Geneva, Switzerland; Glencore of Baar, Switzerland; and Noble Group of Hong Kong;

Vitol, Trafigura and Glencore are said to be the top three crude oil traders in the world.[8] "Vitol and Trafigura sold a combined 8.1 million barrels a day of oil last year (2010), which was equal to the combined oil exports of Saudi Arabia and Venezuela." Koch Industries, Vitol and others in 2009 "parked 100 million barrels of oil in seaborne tankers" for a period of time in order to influence market price.[9]

As with any momentous movement in which there are many competing interests, there is misinformation being promulgated. Announcements by Saudi Arabia, OPEC and the oil industry are rarely questioned by the press. Substituting natural gas for oil is not likely to be in their best interest.

Much of the information we are daily confronted with is false. In the matter of oil reserves, there is no way of verifying what many countries are reporting. T. Boone Pickens says in his book, *The First Billion Is The Hardest*, that he does not believe the amount of the reserves reported by Kuwait, Iran, Iraq, and Saudi Arabia are accurate.[10]

There is a kind of quiet revolution going on in this country as more and more people become aware of the advantages of replacing oil with natural gas as a fuel for transportation. The advantages are it is cheaper, cleaner, causes less pollution, and makes the U.S. less dependent on imported foreign oil.

18

Some say converting to natural gas as a fuel for transportation will enable this country to stop importing foreign oil. On a long term basis, this accomplishment should result in trillions of dollars remaining in circulation in North America rather than being sent overseas.

The U.S. is lagging far behind other countries in converting to natural gas. Companies in the U.S. are producing so much natural gas that the commodity is in oversupply domestically. Therefore they are looking at overseas markets to sell their product since demand is weak in this county. Construction is planned at a number of ports in the U.S. and Canada of facilities to export U.S. natural gas to other countries.

There are 14 countries around the world that have more natural gas vehicles on the road than the United States. This despite the U.S. being fifth in the amount of natural gas reserves (behind Iran, Russia, Qatar and Saudi Arabia) and having the most advanced vehicular transportation system.[11]

Replacing oil with natural gas is complex. There are many companies, agencies and entities either directly or indirectly involved. Entities involved in addition to oil companies include hundreds of federal, state and local government agencies; hundreds of transportation entities such as school and city bus lines, railroads, trucking companies and delivery services; hundreds (even thousands) of private companies; and many non-governmental consulting organizations.

Currently in the U.S., there are possibly hundreds of major and minor oil companies in some way employed in the production, servicing, refining, distribution and retailing of petroleum products and natural gas. Most will be affected by conversion to natural gas.

In addition to the five "supermajor" oil companies listed above, there are some 65 other companies involved in refining petroleum products in the U.S.[12] They operate 141 oil refineries in 31 states.[13] Many are among the hundred or so companies operating rigs that are drilling for natural gas.

Efforts by the private sector are being made to publicize, educate and encourage the adoption of natural gas as a transportation fuel. All of the major players involved in natural gas, including representatives of engine and truck manufacturers, gas companies and trucking fleets attended a major three-day alternative energy convention in Long Beach, California this past May 2012. The Alternative Clean Transportation (ACT) Expo 2012, hosted by Santa Monica, California based consulting firm Gladstein, Neandross & Associates (GNA), attracted hundreds of exhibitors and visitors and was sold out in advance.[14]

There were speakers on how to best utilize natural gas as a transportation fuel and how the refueling infrastructure will be expanded in the near future. Proof-of-concept trucks were on display during the conference for a first hand inspection. In addition, Freightliner initiated a coast-to-coast driving demonstration for a heavy duty truck using natural gas as the sole fuel supply.[15]

END NOTES:
A companion web site has been established to go along with this report so that readers may view pictures, graphs and maps in color and link direct to referenced web sites.

The URL to this "Master" link is:

http://natgas-rpt.com

1 We say "are" manipulated. This is based on the large number of cases involving attempts to manipulate the NYMEX, which are brought every year by the Commodities Futures Trading Commission. Retrieved from **http://www.cftc.gov/index.htm**

2 The CME Group Inc., operates the CME (Chicago Mercantile Exchange), the CBOT (Chicago Board of Trade), the NYMEX (New York Mercantile Exchange), and the COMEX (Commodity Exchange). Retrieved from **http://www.cmegroup.com**

3 The U.S. Commodities Futures Trading Commission regulates commodity markets. Retrieved from **http://www.cftc.gov/index.htm**

4. The list published here was taken from the Energy Information Administration (July 2012) The URL has since been removed. Petro Strategies, Inc. lists 50 of the world's largest oil companies according to size of reserves. Retrieved from **http://www.petrostrategies.org/Links/ worlds_largest_oil_and_gas_companies.htm**

5 The Commodity Futures Trading Commission has a list of violations by British Petroleum (BP). Retrieved from **http://www.cftc.gov/ucm/groups/public/@newsroom/documents/file/p r5405-07_prdoj.pdf**

6 Learsy, R.J. (2012) Oil and Finance, The Epic Corruption Continues 2010-2012. Amazon. (His web site is **http://blog.raymondlearsy.com**)

7 Schneyer, J. (2011, October 28) Commodity traders: The trillion dollar club. *Reuters*. Retrieved from **http://www.reuters.com/article/2011/10/28/us-com modities-houses-idUSTRE79R4S320111028**

8 *Reuters*, Top ten global oil and commodities traders. (2011, April 20), *The Telegraph*. Retrieved from **http://www.telegraph.co.uk/finance/commodities/8451455/Top-ten- global-oil-and-commodities-traders.html**

9 Schneyer. J. op. cit.

10 Pickens, T. B. (2009) *The First Billion Is the Hardest*, New York: Random House

11 Current Natural Gas Vehicles Statistics (2011, December) NGV Global, Natural Gas Vehicle Knowledge Base. Retrieved from **http://www.iangv.org/current-ngv-stats/**

12 Also listed among the supermajor oil companies is Total SA, based in Paris, France. It has a lesser presence in the U.S. than the other five supermajor's. It operates in the U.S. under the name Total Petrochemicals & Refining USA, Inc. Retrieved from **http://www.totalpetrochemicalsusa.com/**

13 Following are the number of refineries in each state and the barrels of oil they are capable of refining on a daily basis. These figures were extracted from U.S. Energy Information Administration (June 2012)

Alabama	3	120,100	Nevada	1	2,000
Alaska	6	389,980	New Jersey	6	715,000
Arkansas	2	82,500	New Mexico	3	142,900
California	16	1,694,621	North Dakota	1	58,000
Colorado	2	102,000	Ohio	3	384,000
Delaware	1	182,200	Oklahoma	6	508,700
Georgia	1	28,000	Pennsylvania	5	773,000
Hawaii	2	147,500	Tennessee	2	362,200
Illinois	4	974,100	Texas	20	3,747,530
Indiana	2	431,500	Utah	5	167,700
Kansas	2	223,000	Virginia	1	66,300
Kentucky	2	217,500	Washington	4	507,850
Louisiana	16	3,492,290	West Virginia	1	20,000
Michigan	1	106,000	Wisconsin	1	34,300
Minnesota	2	336,000	Wyoming	6	166,100
Mississippi	3	365,500	----------------	---	------------
Montana	4	187,600	Totals	133	16,553,771

14 Alternative Clean Transportation (ACT) Expo 2012, Long Beach, California, May 15-17, 2012. Retrieved from **http://www.actexpo.com/**

15 LA to DC on CNG, The Freightliner National Gas Tour. Retrieved from **https://www.freightlinerhaulerchallenge.com/app/facebook**

CHAPTER II
AMERICA'S PROBLEMS

America has numerous problems that are large in scope and complex in nature. They range from a crushing deficit and debt, and a high unemployment rate to depressed real estate prices, two foreign wars and many more issues in between such as funding terrorism and reducing pollution. However, there is one significant problem that can be resolved. That is America's dependence on foreign oil and the transfer of hundreds of billions of dollars to the Middle East.

Solving this problem can help alleviate other issues related to the economy. Many jobs in the near future could be gained as a direct and indirect result of switching to natural gas.

In addition, there will be advancements in personal health by decreasing exhaust from diesel and gasoline fuels. The International Agency for Research on Cancer—part of the World Health Organization—announced at a meeting in France that diesel exhaust is a cause of lung cancer, and has a positive association with an increased risk of bladder cancer.[1-2]

The United States has been addicted to oil as a transportation fuel for decades which has seen hundreds of billions of dollars every year go to foreign countries instead of circulating within North America.

This nation should be well on its way to quitting its reliance on foreign oil, but it is not. The solution is certainly possible now. The U.S. can substitute natural gas for diesel in the operation of medium and heavy duty trucks. Steps have been taken by some to do this. But these steps pale in comparison to what could be done.

By "foreign imported oil," we mean oil originating in countries outside of North America. Canada and Mexico are viewed as peaceful neighbors who share a long common border and are all part of the North American Free Trade Agreement (NAFTA). Our economies and people are so closely linked that the flow of money travels in every direction. Therefore our definition of "foreign imported oil" does not include oil imported from these two countries and in this treatise will be considered as "foreign oil."

THE SAD TRUTH

The sad truth is that America has been at odds with itself in coordinating efforts to achieve oil independence which has resulted in a slow rate in the adoption of natural gas. However in just a few years, this country is expected to start exporting liquefied natural gas (LNG) around the world especially to Asia.

The U.S. could have the worst of all worlds. America will remain addicted to "foreign imported oil" and continue to send many billions of dollars overseas. Prices won't go lower since demand for oil will exceed supply in the U.S. and the rest of the world.

Meanwhile the U.S. will begin exporting LNG as a transportation fuel which this country should first be using to meet its domestic needs for achieving energy independence.

The major use of oil goes towards powering transportation, specifically more than 260 million automobiles and trucks as of 2010.[3] Progress has occurred in increasing the mileage per gallon in automobiles by developing hybrid engines, decreasing the size of cars and creating better aerodynamic designs.

The United States had a generous supply of cheap oil for much of the 19th and 20th centuries. After World War II conditions began to change. By 1973 the United States received about one third of its oil from overseas and a convergence of events developed that shocked the nation.

OPEC

The Organization of Petroleum Exporting Countries (OPEC) was rising in stature and power when another war in the Middle East erupted. This time, OPEC responded with a form of economic warfare by cutting off its petroleum exports to America. The oil boycott damaged our economy as it resulted in gas rationing and lines at the gas pumps. Since that crisis, every president, Congress and Americans in general have pledged to get us off of dependency on foreign oil.

In 2005 this country was importing about two thirds of its oil which was twice the percentage when the first oil crisis occurred in 1973. By 2012, America had achieved a degree of progress through high -mileage cars like those outfitted with hybrid car engines, but dependency on foreign oil and a threat of a boycott for political and military reasons remains.

The current deep recession has also resulted in a decreasing amount of oil consumption due to lower economic activity. However as soon as business improves, oil consumption will rise again. The bottom line is after 39 years from our wake-up call in 1973, the United States is still begging foreign countries for their oil.

END NOTES:
A companion web site has been established to go along with this report so that readers may view pictures, graphs and maps in color and link direct to referenced web sites.

The URL to this "Master" link is:

http://natgas-rpt.com

1 Cart, J, (2012, June 12) Diesel engine exhaust linked to increased risk of lung cancer. *Los Angeles Times.* Retrieved from **http://www.latimes.com/news/science/sciencenow/la-sci-sn-diesel-engine-exhaust-linked-with-risk-of-lung-cancer-20120612,0,7726895.story_**

2 Diesel engine exhaust carcinogenic (2012, June 12,) International Agency for Research on Cancer (IARC) World Health Organization. Retrieved from_**http://press.iarc.fr/pr213_E.pdf**

3. Table 1-11: Number of U.S. aircraft, vehicles, vessels and other conveyances, RITA, Research and Innovative Technology Administration, Bureau of Transportation Statistics. Retrieved from **http://www.bts.gov/publications /national_transportation_statistics/html/table_01_11.html**

CHAPTER III
TERRORISM AND WAR

There have been ongoing terrorist activities for many decades often emanating from the Middle East where much of our oil imports originate. Hundreds of billions of American dollars are sent to the Middle East every year where there are those who siphon off money to finance terrorist organizations. The access to this money by radical Islamists and other types of terrorists supports their operations.

The cost in human lives and money spent in upholding our oil imports is enormous. Every American knows what happened on September 11, 2001 when passenger planes were used as bombs to destroy the twin towers of the World Trade Center.

Controversy exists over whether or not bin Laden's family wealth directly financed the 9/11 event as well as Al-Qaeda operations in general. The bin Laden's family patriarch rose to high status and he was a very wealthy individual through his extensive construction and business activities. The father was close to the ruling family in Saudi Arabia and unexpectedly died in an airplane crash in 1967 leaving many heirs to his fortune. How much money if any bin Laden received remains a mystery but he had access to substantial funds. He personally had raised money in the early years of struggle against the Russians who had invaded Afghanistan.

Bin Laden had gained enough financial resources to build a number of camps in that country. He was "officially" cut off from the family's fortune during the early 1990's and since then his personal finances can't be reliably substantiated.[1]

There is a money exchanging process in the Middle East called Hawala that is an ancient process of transferring funds on a handshake through an understanding of trust.[2] On many levels, Hawala has been a very beneficial mechanism in funding worthwhile social welfare activities. However there is a dark side of this practice which needs to be understood and separated from the real charitable offerings.

There is often no formal paper trail or official documentation that is traceable via modern telecommunications and computerized systems. This is one reason for its popularity in today's world for supporting radical organizations and terrorism.[3] The leakage of oil derived income to pay for terrorist organizations is essential to the ongoing financing of terrorism. In essence, the United States paid on an indirect basis for the 9/11 terrorist bombings that killed 2,752 of its citizens.

PAYING FOR TERRORISM

According to James Woolsey, former director of the Central Intelligence Agency, "Except for our own Civil War, this [the war on terror] is the only war we fought where we are paying for both sides. We pay Saudi Arabia $160 billion for its oil, and $3 or $4 billion of that goes to the Wahhabis, who teach children to hate. We are paying for these terrorists with our SUV's."[4]

New technology in developing natural gas (especially shale gas) and in building natural gas engines for heavy duty trucks enables America to quickly reduce its oil imports. But this is not happening in rapid fashion.

Rather the opposite is occurring in that obstacles by the power elite are blocking conversion to natural gas as a transportation fuel which is preventing America from stopping the subsidy of terrorism with its own money.

Historical descriptions of how America has extended its military reach, sacrificed its troops and spent untold amounts of money can go back many decades. Protecting countries, from which foreign oil is imported, is a major consideration in how the U.S. deploys its military.

Saddam Hussein invaded Kuwait in 1991. Under the auspices of the United Nations, President George H.W. Bush organized a coalition of countries which drove Saddam out of Kuwait. The war was a success as measured by the withdrawal of the Iraqis from Kuwait and by the general degradation of Saddam's army. Saddam remained in power but was restricted in his political and military actions. One of the considerations for the first Gulf War was to protect America's oil interests in Kuwait and Saudi Arabia.

On September 11, 2001, two passenger planes loaded with fuel crashed into the twin towers of the World Trade Center in New York. In response, the U.S. invaded Afghanistan and drove out Al-Qaeda and bin Laden, who claimed responsibility for the attack.

In March 2003, George W. Bush opened a second front with the invasion of Iraq. There is considerable and continuing controversy over the reason for the invasion. Ostensibly it was because Saddam Hussein was developing and stock-piling nuclear and chemical weapons of mass destruction and was collaborating with bin Laden. As it turned out there were no weapons of mass destruction and no collaboration with bin Laden.

Many have backed the position that the invasion of Iraq was motivated by oil. They include Alan Greenspan, former chairman of the U.S. Federal Reserve Board. Prior to the

Iraqi war, there was much discussion by members of the Bush Administration, ministers of the UK Labor government and some of the world's largest oil companies, including BP and Shell Oil, about the prospect of developing Iraqi oil fields. Whether oil was the overriding consideration for the invasion will likely never be known. However, there is little doubt that it was a consideration.[5]

The cost of these two wars is staggering. An article in Reuters June 29, 2011, states, "The final bill will run at least $3.7 trillion and could reach as high as $4.4 trillion, according to the research project 'Costs of War' by Brown University's Watson Institute for International Studies."[6]

END NOTES:
A companion web site has been established to go along with this report so that readers may view pictures, graphs and maps in color and link direct to referenced web sites.

The URL to this "Master" link is:

http://natgas-rpt.com

1. A biography of Osama Bin Laden Laden (1999, April and 2001, September 13) Frontline, Public Broadcasting Station **http://www.pbs.org/wgbh/pages/frontline/shows/binladen/who/bio2.html**

2. Greenberg, M.R., Weschsler, W. F. and Wolosky, L. S. (2002)"Terrorist Financing, Report of Independent Task Force" Council on Foreign Relations. **http://www.cfr.org/economics/terrorist-financing/p5080**

3. Kleymeyer, J. (2003) "Hawala: An alternative banking system and its connections to blood diamonds, terrorism, & child soldiers", TED Case Studies, Number 119. **http://www1.american.edu/ted/hawala.htm**

4. Powers, J. (2010, February 7) "Oil Addiction: Fueling Our Enemies", Truman National Security Project. **http://www.trumanproject.org/files/papers/Oil_Addiction__Fueling_Our_Enemies_FINAL.pdf**

5. Adams, R. (2007, September 16) "Invasion of Iraq was driven by oil, says Greenspan", The Guardian.
http://www.guardian.co.uk/world/2007/sep/17/iraq.oil

6. Trotta, D. (2011, June 29) "Cost of war at least $3.7 trillion and counting." **http://www.reuters.com/article/2011/06/29/us-usa-war-idUSTRE75S25320110629**

CHAPTER IV
POLLUTION

The burning of fossil fuel such as oil produces noxious and eventually poisonous particulates and chemicals which affect all living organisms. Along with coal, the burning of oil through gas and diesel products releases carbon dioxide, sulfur dioxide and nitrogen dioxide into the atmosphere which float around for long periods of time. When rainfall mixes with these chemical particulates, it produces what is called acid rain which is very detrimental to plant life.[1] Forests and other vegetation are adversely affected but our food supply is somewhat protected by adding lime and fertilizers that replace lost nutrients.[2]

Smog develops from a combination of factors including "heavy motor vehicle traffic, high temperatures, sunshine, and calm winds. Weather and geography affect the location and severity of smog. Because temperature regulates the length of time it takes for smog to form, smog can occur more quickly and be more severe on a hot, sunny day."[3]

There is a mixture of air pollutants that include the nitrogen oxides previously listed and volatile organic compounds which form ozone. Ozone high in the atmosphere is beneficial, since it protects our planet from high doses of radiation, but ground level ozone creates smog[4].

Our eyes burn, breathing is more difficult and people with medical problems such as cardiac conditions and asthma are at risk for becoming more ill. Increased hospitalizations often occur on days of intense smog.[5]

A new terminology has been developed that describes various levels of smog alerts that necessitate a reduction in physical activity for school children as well as the public. Standardized measurements by the US Environmental Protection Agency (EPA) have resulted in the Air Quality Index[6]. Cities now report to the public and keep records of their performance in keeping the air we breathe clean and healthy. The reduction in exhaust from all types of vehicles has been a major goal for decades.

POLLUTION AND CLIMATE CHANGE

Gas and diesel fuel are all very dirty pollutants and require extensive smog control devices which are expensive to install and to maintain on a long term basis. Natural gas on the other hand is a significant improvement. "Natural gas does not contribute significantly to smog formation, as it emits low levels of nitrogen oxides, and virtually no particulate matter. For this reason, it can be used to help combat smog formation in those areas where ground level air quality is poor."[7] If we can continue to develop sufficient supplies of natural gas and safe and effective methods to drill for it, the country will be better off.

Evidence indicates that human activity primarily through the burning of fossil fuels has resulted in higher temperatures worldwide. This has affected weather patterns, plant growth, animal adaptation to their habitat and food crop production to name a few consequences. [8]

Glaciers at many points in the world are receding and deserts are expanding since areas of the world that are exposed to the more extreme weather, the coldest and hottest lands, are affected the most in the beginning stages of global warming.[9]

It has been difficult to discern the differences between normal variations in weather conditions vs. long term trends in global warming. The ability to distinguish the fluctuating background noise in temperatures and environmental changes and what are long term trends due to global warming has been a complex process.[10]

There is a great deal of controversy about global warming, whether it exists or not. Much of the denial of it is politically motivated. This treatise takes the position that the science of it is valid.

The scientific community has determined that the rate of temperature change has accelerated greatly in the last 50 years. In fact, the current levels of carbon dioxide in the atmosphere are higher than at any time during the last 650,000 years. In addition, the world's oceans have absorbed about 20 times as much heat as the atmosphere over the past half-century, leading to higher temperatures not only in surface waters but also in water 1,500 feet below the surface. The measured increases in water temperature lie well outside the bounds of natural climate variation.[11]

Union of Concerned Scientists stated that U.S. passenger vehicles are one of the largest contributors to global pollution, responsible for roughly a quarter of annual U.S. energy-related emissions of CO_2.[12]

Global warming denial has had a number of supporters whose decisions have mostly been based on a political philosophy rather than a scientific one. Often their approach emphasizes that government is the source of information on the matter and that there is a divided scientific community.

The government hasn't originated the premises underlying global warming since the data is generated from the scientific community. Government may use the information to establish and implement policy but it doesn't generate the facts behind the claims of global warming.

The deniers also propagate the perception that there is a divided scientific community in establishing the fact of global warming. In actual fact the contrary is true. A study in 2009 reviewed an extensive dataset of 1,732 climate researchers and their publications. There was a 97 percent to 98 percent agreement that global warming from human activity is valid. There is also substantial consensus among scientists in regards to the effects of a warmer planet.[13]

An important consideration is that there is only one atmosphere encompassing the earth. There will be losers and winners as the climate changes due to global warming. Some lands and peoples will benefit while others will not. It is impossible to predict exactly what will occur on a long term basis.

For example, long term droughts worldwide have persisted and threaten people's current way of life.[14] The conditions in Texas are just one example of a territory grappling with major economic consequences not only locally but perhaps globally too.[15, 16]

CONFLICTS

The lack of water has created international conflicts between Mexico and the United States (i.e. Texas) over when and how much water should be released along the Rio Grande River which separates the two countries.[17] It appears that climate science experts aren't very optimistic that conditions will improve but in fact predict that the drought will intensify in the future.[18]

Texas is in a huge dilemma since large swaths of its lands don't have water which in turn is a consequence of global warming. On the other hand, there are other powerful economic and political segments of the state which support vast drilling and consumption of oil and natural gas and even the mining for uranium. It is as if Texas is a micro-chasm of the worse effects of higher temperatures and in contrast the beneficial effects of depending on drilling and consumption of fossil fuels.

CONFLICT OVER WATER

One has to keep in mind that great amounts of water are often used in the drilling and mining industries, so there could be additional conflicts even in the oil rich lands over dwindling water resources. Much of the water in Texas originates from large underground aquifers[19] There is disagreement over how long it takes for the aquifers to be replenished especially during extended drought conditions. Will the state become a waste land in one area and wealthy in another? Will Texas ignore the effects of global warming and continue on its current course indefinitely towards an uncertain future?

In contrast to Texas that has many resources at its disposal in handling drought conditions, Africa has been unable to cope with its own dwindling rainfall. Famine has killed many people and persists with calamitous results as governments have been ineffective in confronting this disaster. The leadership in many countries is either incompetent, unstable or engages in war which makes the situation worse.

What transpires in one part of our world often affects events elsewhere. Earth has become a smaller planet in that trans-portation, telecommunications and economic linkages enable people and information to travel faster than ever before. Political consequences are felt worldwide.

The famine in Africa could lead to the establishment of extremist and totalitarian governments as a response to the disaster from drought. Fighting in Eritrea and elsewhere is by Islamic extremists who take advantage of government failures to provide for their people. The world, especially America, is well aware of what occurred last time when a radical Islamic government in Afghanistan gave cover to Al-Qaeda.

Are there any benefits from rising temperatures and differing amounts of rainfall or melting snow? Canada could be the beneficiary of establishing a Northwest Passage if there is continued melting of its snow pack and Tundra. Shipping between the eastern and western hemispheres would become much more efficient although the new lanes probably would not be operating year round.[20] This would become an economic boon to Canada and trading partners in the East and West. However these positive outcomes could be tempered to some extent by release of methane trapped within the frozen tundra which would harm our ozone layer.[21]

The controversial climate warming issue brings out emotional and political claims that do not have anything to do with the science or outcomes of what is occurring. The bottom line is that natural gas as a transportation fuel is cleaner for our earth in comparison to petroleum products. Therefore the switch to natural gas will improve the health of the planet and for life in general. This is fact and should be recognized as one of the benefits of converting to natural gas.

END NOTES:
A companion web site has been established to go along with this report so that readers may view pictures, graphs and maps in color and link direct to referenced web sites.

The URL to this "Master" link is:

http://natgas-rpt.com

1 Oldroyd, C., What can I do to decrease acid rain? National Geographic, Green Living, *Demand Media.* Retrieved **http://green livingnationalgeographic.com/can-decrease-acid-rain-2618.html**

2 What is acid rain? (2007, June 8) U. S. Environmental Protection Agency (EPA). Retrieved from **http://www.epa. gov/acidrain/what/index.html**

3 West, L., What causes smog? *About.com, Environmental* Issues. Retrieved from **http://environment.about.com/od/smogfaq/f/ smog_faq_five.htm**

4 Good up high bad nearby – What is ozone? *AIR Now* (product of U.S. EPA NOAA, NPS, tribal, state, and local agencies). Retrieved from **http://www.airnow.gov/index.cfm?action=goodup.page1**

5 West, L., What are the effects of smog? *About.com, Environmental Issues*. Retrieved from **http://environment. about.com/od/smogfaq/f/smog_faq_three.htm**

6 Air Quality Index (AQI) – A Guide to Air Quality and your Health. (2011, December 9) *AIR Now* (product of U.S. EPA NOAA, NPS, tribal, state, and local agencies). Retrieved from **http://www.airnow.gov/index.cfm?action=aqibasics.aqi**

7 Human Fingerprints, Global Warming, (2006, May 11) Union of Concerned Scientists. Retrieved from_**http://www.ucsusa.org/ global_warming/science_and_impacts/science/global-warming-human.html**

8 Human Fingerprints, Global Warming, Ibid.

9 Pelto, M. S., Recent Global Glacier Retreat Overview. North Glacier Climate Project. Retrieved from **http://www.nichols.edu/departments/glacier/glacier_retreat.htm**

10 Human Fingerprints, Global Warming, op cit.

11 Human Fingerprints, Global Warming , op.cit.

12 Vehicle Solutions, Global Warming. Union of Concerned Scientists. Retrieved from **http://www.ucsusa.org/global_warming/solutions/vehicle_solutions**

13 Anderegg, W.R.L., Prall, J. W., Harold, J. & Schneider, S. H. (2010, April 9) Expert credibility in climate change. Proceedings of the National Academy of Sciences of the United States of America (PNAS). Retrieved from **http://www.pnas.org/content/ 107/27/12107.full**

14 Global Drought Monitor (2012, May) University College London (UC), Department of Space and Climate Physics. **http://drought.mssl.ucl.ac.uk/drought.html?map=%2Fwww%2Fdro ught%2Fweb_pages%2Fdrought.map&program=%2Fcgi-bin%2Fmapserv&root=%2Fwww%2Fdrought2%2F&map_web_im agepath=%2Ftmp%2F&map_web_imageurl=%2Ftmp%2F&map_w eb_template=%2Fdrought.html**

15 Texas Drought (2012, August 8 updated) Drought remains as lakes
are about half full. Lower Colorado River Authority, (LCRA). Retrieved
from **http://www.lcra.org/water/drought/index.html**

16 Galbraith, K. (2011, October 30) Catastrophic drought in Texas
causes global economic ripples. *The New York Times.* Retrieved from
**http://www.nytimes.com/2011/10/31/
business/energy-environment/catastrophic-drought-in-texas-causes-
global-economic-ripples.html?_r=4**

17 Sherman, C. (2012, April 20) Drought sparks water dispute between
Texas and Mexico, Irrigation divides farmers on both sides of border.
AZCenteral.com., *The Associated Press.* Retrieved from
**http://www.azcentral.com/news/articles/2012/04/
19/20120419drought-sparks-water-dispute-texas-mexico.html**

18 Wagner, S. (2012, February 1) Climate science experts predict
intensified drought in Texas. *chron.com.* Retrieved from
**http://blog.chron.com/txpotomac/2012/02/climate-science-experts-
predict-intensified-drought-in-texas/**

19 Baker, B. B., Underground water, Handbook of Texas Online. Texas
State Historical Association. Retrieved from
http://www.tshaonline.org/handbook/online/articles/gru01

20 King, H., Northwest Passage – Map of arctic sea ice, Global warming
is opening Canada's Arctic. *geology.com.* Retrieved from
http://geology.com/articles/northwest-passage.shtml

21 Barnett, A. (2008, December 3) Methane bursts from frozen tundra.
Ice build up may squeeze greenhouse gas from cold soil, *Nature,
International weekly journal of science.* Retrieved from
http://www.nature.com/news/2008/081203/full/news.2008.1275.html

CHAPTER V
THE TRANSFER OF WEALTH

In general, the U.S. has relied on increasingly larger amounts of imported oil which has resulted in probably the most massive transfer of wealth in the history of the world. It is probably impossible to find agreement on a definite figure, but it would be safe to say it would come to many trillions of dollars. A brief summary of the fluctuations in oil imports was explained in the Introduction of a report by Congressional Research Service that used data from Energy Information Administration.

"Imports have generally increased over the last six decades, except for a period following the oil spikes of the 1970s and again in the last five years. Oil import volumes peaked in 2005 and then declined through 2010 as a result of economic and policy driven changes in domestic supply and demand. However, oil import costs have increased due to rising prices, which more than offset the savings from lower import volumes."[1]

Highlights of the above report indicated the following changes in the percentages of oil imports by U.S:

1973 during oil embargo	30%
2000	50%
2005	more than 60%
2011	less than 50%.[2]

How much do Americans pay for imported oil and what are the economic effects of this outlay? How does the global economy and demand for oil around the world affect the pocketbooks of Americans? These are important issues especially in these times of high unemployment and high cumulative deficits.

T. Boone Pickens developed a strategy which he labeled the Pickens Plan to reduce imported oil in America. Originally he wanted to reduce reliance on natural gas to generate electricity by constructing wind turbines and connecting them into the national energy grid. The freed up natural gas as well as increased production of this resource from shale formations would then be allocated as a transportation fuel to replace gasoline and diesel.

Pickens' current plan is to delay the wind turbine portion of the strategy and rely solely on natural gas shale deposits for use as a transportation fuel for trucks. Apparently there are substantial amounts of natural gas to support its application for vehicles and the cost/benefits of the wind turbines became impractical.

Estimates of the monthly and yearly amounts of money transferred for payment of imported oil into the United States are astounding.. Pickens' web site entitled PickensPlan.org states that during June 2012 the U.S. imported 332 million barrels of oil from all sources at a cost of $32.2 billion which is approximately $745,000 per minute. What would happen if these dollars could be re-circulated within the United States? It would mean greater prosperity, lower unemployment and improved independence from nations who want to harm the U.S.[3]

The definition of imported oil used by Pickens and the Energy Information Administration is to lump together the costs of oil from all sources outside of the U.S including Canada and Mexico.

In this treatise we make the distinction between "imported oil" and "imported foreign oil." The reason is Canada and Mexico are the two largest sources of imported oil and there is no energy security threat from either country. We have peaceful partnerships and share common borders with both countries. Therefore, the most realistic and practical economic objective involves eliminating oil from the Middle East and other radicalized countries. This is where energy security is a major concern as well as not wanting to support terrorist organizations and radical Islam.

The EIA has a listing of imported oil by country of origin over a six-year period.[4] According to the EIA, the eight largest exporters of oil outside of North America to the U.S. in 2011 were:

Saudi Arabia	436.051,000
Venezuela	344,685,000
Nigeria	298,153,000
Russia	226,675,000
Iraq	167,905,000
Columbia	154,170,000
Algeria	130,568,000
Angola	126,259,000

All were members of OPEC except Russia and Columbia. Canada and Mexico are the two largest exporters of oil to the U.S. Their 2011 exports to the U.S. were:

Canada	987,736,000
Mexico	439,753,000

The big oil companies, their suppliers and servicing companies are enjoying high demand for fossil fuel whether it's imported or not. They make money on the upside in assisting foreign countries to drill and process energy and they make even more money at the down side in the United States through refining, distribution and sale of oil and petroleum products. They aren't interested in getting off of oil since their livelihood is based on receiving the highest possible net profit.

NO FREE MARKET

Chairman of Federal Express Frederick Smith stated in an interview that "There is no free market for oil. It's controlled by a cartel, OPEC ... They own 80 percent to 90 percent of the reserves in the world. They produce about 42 percent." He continued to say that the principal problem "... is about energy conservation and national security by getting us off of imported oil." He claimed that the cost of imported oil from all sources has harmed the average family economically since their expenses increased from an estimated $1,700 in 2001 to $4,000 in 2011.[5]

On a historic basis during the 1970's and 1980's, there was resistance to improving gas mileage of cars by the auto industry. One gallon of conservation is equivalent to eliminating one gallon of imported oil which is just as important as drilling for more fuel. Both strategies are equally important for the United States to become more energy independent.

Corporate Average Fuel Economy (CAFE) rules have been established through the joint efforts by industry and government to provide objective data for consumers when they purchase vehicles. This information will enable buyers to select high mileage cars based on valid research.

Recent agreements have resulted in establishing new mileage goals that will increase until 2025.[6]

The advances in fuel consumption have been possible principally through reducing the size of automobiles, development of hybrid engines and use of computerized powertrain functions.

These higher standards are probably one of the most significant victories in reducing oil imports that has occurred in many decades. The only issue is that this process will take a long period of time to achieve its stated goal since 2025 is still 12 years away. Also improving gas mileage for passenger cars may not be sufficient in getting America off of imported foreign oil. There has to be some kind of effective action which can be achieved in conjunction with improved gas mileage, and that answer lies in natural gas as a transportation fuel.

JOBS CREATED

Hundreds of thousands of direct and indirect jobs will be created in the future if shale reserves of natural gas are drilled.[7] This additional fuel would then be directed for transportation, manufacturing and energy generating pur-poses which become an employment engine for America. There has to be a valid long term application to economically use the new-found volume of natural gas that is being extracted within the United States.

Businesses would save billions of dollars over the years in fuel costs, which will translate into more profitable companies and maybe lower gas prices for consumers. Tax receipts would increase on local, state and federal levels given the additional number of people with jobs. For example, an online article describing one study by the Associated Press indicated that natural gas drilling in 2011 produced revenues of approximately $3.5 billion in Pennsyl-vania and $2.1 billion in West Virginia. These amounts have the potential of growing substantially if current depressed price levels of natural gas increase in the future.[8]

It was also estimated that even more money and jobs would be earned by using natural gas in the chemical industries to produce industrial and other compounds. For example, ethylene is one of the highest volume chemical products in the world and it can be derived either from higher cost oil or from much lower cost natural gas. It is the basis of plastic products and other organic- based goods like fertilizer. A number of new ethylene chemical plants will be constructed in U.S. in the near future as a result of the lower cost of natural gas.[9-11]

The domestic oil industry and refining companies will be impacted but they have the rest of the world to do business with. The chemical companies will do extremely well as outlined in the preceding statement by using natural gas as a base to manufacture their products instead of the more expensive and polluting petroleum.

There are sufficient supplies to support the application of natural gas for both transportation purposes and chemical/manufacturing industries. Fortunately, the U.S. has enormous shale formations and is able to generate methane from biogas sources too. All of these business sectors will benefit and there isn't a mutually exclusive condition of winner vs. loser.

END NOTES:

A companion web site has been established to go along with this report so that readers may view pictures, graphs and maps in color and link direct to referenced web sites.

The URL to this "Master" link is:

http://natgas-rpt.com

1 Nerurkar, N. (2011, April 1) U.S. Oil Imports: Context and Considerations, Congressional Research Service. Retrieved from **http://www.fas.org/sgp/crs/misc/R41765.pdf**

2 U.S. Oil Imports: Context and Considerations, ibid.

3. PickensPlan, (June 2012) Monthly Imports. Retrieved from **http://www.pickensplan.com/oilimports**

4 Petroleum and other liquids, U.S. imports by country of origin, Energy Information Administration (EIA). Retrieved from: **http://www.eia.gov/dnav/pet/pet_move_impcus_a2_nus_ep00_im0 _mbbl_a.htm**

5 Beller, M. D. (2012, April 16) US needs plan to cut OPEC oil imports: FedEx CEO, *CNBC*. Retrieved from **http://www.cnbc.com/id/47062294**

6 NHTSA and EPA Propose to Extend the National Program to Improve Fuel Economy and Greenhouse Gases for Passenger Cars and Light Trucks, (2011, November), Office of Transportation and Air Quality Environmental Protection Agency, Regulatory announcement. Retrieved from **http://www.epa.gov/oms/climate/documents/420f11038.pdf**

7 Energy production: the key to prosperity, (2012, May 3) Louisiana Mid-Continent Oil and Gas Association (LMOGA). Retrieved from **http://www.lmoga.com/news/jack-gerard-energy-production-the-key-to-prosperity**

8 Begos,K. (2012, May 5) Pa. gas drilling brought $3.5 billion in 2011, *Bloomberg News*. Retrieved from **http://www.businessweek.com/ap/2012-05/D9UIKVN00.htm**

9 Kaskey, J. (2012, April 9) Dow to build ethylene plant in Texas on cheap gas prices, *Bloomberg News*. Retrieved from **http://www.bloomberg.com/news/2012-04-19/dow-to-build-ethylene-plant-in-texas-on-cheap-gas-prices.**

10 NPR Staff and Wires, (2012, March 16) Shell picks Pittsburgh area for major refinery, *National Public Radio*. **http://www.npr.org/2012/03/16/148702640/shell-picks-pittsburgh-area-for-major-refinery**

11 Carter, J. (2011, February 23) Why ethylene is so important to chemical industry, *Helium*. Retrieved from **http://www.helium.com/items/2099858-why-ethylene-is-so-important-to-the-chemical-industry**

48

CHAPTER VI
NATURAL GAS DISCOVERIES

The United States has been blessed in the past decade or two in that vast new shale natural gas fields or plays have been discovered and explored within our borders. The new finds increased the proven reserves of natural gas to 273 trillion cubic feet as of January 1, 2011 which is the fifth largest worldwide. Only Russia, Iran, Qatar and Saudi Arabia have more proven reserves in the world.[1] In fact, there is less than a one percent difference (two trillion cubic feet) between Saudi Arabia and the U.S.

American ingenuity has invented and developed new methods for recovering the expansive deposits of natural gas and oil in shale formations. The technologies include horizontal drilling and fracturing rock formations which have greatly increased yields[2] The supply of natural gas has expanded so much that the price is at historically low levels. The U.S. has become the acknowledged global leader in applying these new processes.[3]

The old fashion method of drilling involves creating a vertical shaft to tap oil and natural gas reserves. This can be a hit or miss proposition. By being able to turn the drill bits by 90 degrees, drilling companies can recover greater amounts of fuel that run laterally.

MULTIPLE DRILLING AT VARIOUS DEPTHS

There are even methods to have multiple horizontal drilling at various depths depending on the size and shape of the formations. A well can be likened to a trunk of a tree that branches out in directions, which follow the source of oil and gas formations.

At this point in time, fracturing loosens up the natural gas and/or oil that are soaked within the shale deposits. The traditional method of fracturing which is now labeled as "fracking" consists of pouring large amounts of water combined with secret chemical formulas under great amounts of pressure which open the cracks in shale.

The pressurized process has companies constructing larger compressors and at increased levels of horsepower to open the shale over longer lengths of horizontal branching. Sand is included in this mixture which keeps the cracks open when the water is withdrawn. Chesapeake Bay, a major natural gas company, stated that it averages about 4.5 million gallons of water for each fracking treatment.[4] Statements from other sources indicated that three million gallons is about the bottom limit for fracking treatments.

MULTIPLE LAYERS

In some cases multiple layers of shale oil have been discovered at varying depths so that the yield becomes greatly magnified. For example Marcellus Shale is the first layer of shale oil and that is a huge play that stretches across many states in the Northeast and Appalachia.[5] A second layer called Utica Shale is thicker than the first one and appears to be a feasible play as well. It is below the Marcellus Shale.

Unlike oil, natural gas is more widespread throughout the United States rather than clustered in several principle locations These formations range through multiple states in the east to northeast.[6] In fact, 36 major shale deposits have been discovered covering 24 states.[7] The population in these 24 states represents 57 percent of the people in America, a majority of the country's citizens who will benefit immediately from their state's economic expansion into natural gas.[8]

The dispersal of natural gas formations to so many regions throughout America means that transporting the resource to the market place for consumption will be easier and the distances shorter especially in comparison to oil. Natural gas is not centrally located in a few locations where it has to be shipped thousands of miles to local markets.

DRAMATIC INCREASE

The expansion of drilling for natural gas in major oil plays has increased dramatically in recent years. The increase in drilling of natural gas wells in Pennsylvania over a five-year period from 2007 to 2011 is portrayed below and illustrates the boom times of what is occurring in many locations throughout the United States.[9]

Marcellus Shale wells drilled in Pennsylvania per calendar year:[10]

2007   27
2008   161
2009   785
2010   1386
2011   2073

Worldwide there have been enormous shale oil discoveries in countries like Argentina, China and Saudi Arabia. For instance, Argentina's Vaca Muerta natural gas field was recently discovered about two years ago and was considered one of the biggest shale oil finds outside of North America.[11]

The size of the natural gas play will have a great impact on the country. Additional natural gas has been found off of the Falkland Islands.[12]

These new discoveries will be a game changer to the energy business where dirty and higher cost petroleum can be replaced with cleaner burning and lower cost natural gas. Many countries are transferring their transportation fuel to include natural gas. Many countries are ahead of the U.S. in converting to natural gas as a fuel for transportation.

DISPARITY IN COST

The cost of natural gas in the United States has been low. It averaged $2.52 for February 2012. According to World Bank commodity pricing there was considerable difference elsewhere in the world. In Japan the price was at $16.25 and at $11.12 in Europe for the same time period. [13]

America can take advantage of lower manufacturing costs in the chemical industries which use either petroleum or natural gas as a base to create their products A 25% increase in the production of ethane will result in 400,000 more jobs in the country.[14] There should be a corresponding increase in GDP and taxes for the government. One significant issue is whether or not the changes occur at a slow or rapid rate.

Other countries are adopting natural gas as a transportation fuel quicker than the U.S. This will make their trucking industries more cost effective. America is a lowly number 15 in possessing vehicles that operate on natural gas.

Natural gas is about $2 less expensive than diesel and gas depending on the location. The U.S. Energy Information Administration projects this $2 disparity to last through 2035.[15]

As mentioned previously, the chemical industry could provide more than 400,000 jobs through the construction of facilities, hiring of long term personnel and increased employment through indirect economic benefits.[16] NaturalGas.org reported that hundreds of thousands even millions of new jobs could be created by the development of natural gas resources and its application as a transportation fuel. These may be high or over optimistic estimates of employment projections, but the trends are positive in reducing unemployment.[17]

On the other hand, America could experience a worse case scenario. There has been systematic blockage in using natural gas for transportation purposes. Elements in Congress, mainly Republicans, have prevented a partnership between the public and private sectors in fully developing the advantages of natural gas.

Plans for facilities to export natural gas to other nations have been approved by the Federal Energy Regulatory Commission (FERC). Construction on one by Cheniere Enenrgy is expected to be completed in four years.[18, 19]

This country will be faced with higher gasoline prices and still dependent on foreign imported oil for its economic survival. Meanwhile natural gas which can help America become more energy independent will be shipped overseas for other countries to benefit.

END NOTES:
A companion web site has been established to go along with this report so that readers may view pictures, graphs and maps in color and link direct to referenced web sites.

The URL to this "Master" link is:

http://natgas-rpt.com

1. World natural gas reserves by country as of January 1, 2011, (2011) International Energy Outlook 2011, U.S. Energy Information Agency (EIA), p 64, Table 7. Retrieved at **http://www.eia.gov/forecasts/ieo/pdf/0484(2011).pdf**

2 Horizontal – Directional Oil & Gas Well Drilling. Horizontal Drillling.org. Retrieved from **http://www.horizontaldrilling.org/**

3 Horizontal – Directional Oil & Gas Well Drilling, Ibid.

4 Water usage, Hydraulic Fracturing Facts. Chesapeake Energy. Retrieved from **http://www.hydraulicfracturing.com/Water-Usage/Pages/Information.aspx**

5 King, H. Super giant field in the Appalachians? Marcellus Shale - Appalachian Bay Natural Gas Play. *geology.com*. Retrieved from **http://geology.com/articles/marcellus-shale.shtml**

6 Map of shale gas basins in the United States. EnergyindustryPhotos.com. Retrieved from primary source located in U.S. Department of Energy, Energy Information Administration (EIA). Retrieved from **http://energyindustry photos.com/shale_gas_map_shale_basins.htm**

7 North American shale fields - Natural gas and oil shale fields in North America, United States – USA shale fields. Retrieved from **http://oilsh alegas.com/shalefields.html**

8 Population Density, Resident Population Data. United States Census 2010. Retrieved **from http://2010.census.gov/2010 census/data/apportionment-dens-text.php**

9 King, H. Super giant field in the Appalachians?, op.cit.

10 King, H. Super giant field in the Appalachians?, op.cit

11 Daly, J. (2011, November 10) Argentina announces massive oil and natural gas reserves. *OILPRICE.com*. Retrieved from **http://oilprice.com/Energy/Energy-General/Argentina-Announces-Massive-Oil-And-Natural-Gas-Reserves.html**

12 Young, S. (2012, April 23) Large Falklands natural gas discovery will raise tensions with Argentina. Langley Intelligence Group Network. Retrieved from **http://www.lignet.com/Wire/UPDATE-2-Borders---Southern-Makes-Falklands-Gas-Di**

13 Commodity Price Data, World Bank. (2012, March 5)   Retrieved from **http://siteresources.worldbank.org/INTPROSPECTS/Resour ces/33 4934-1111002388669/829392-1325803576657/Pnk_0312.pdf**

14 Shale gas and new petrochemicals investment: Benefits for the economy, jobs and US manufacturing. (2011, March) Economic & Statistics, American Chemistry Council, Executive summary and p.26, Tables 2 & 3. Retrieved from **http://www.americanch emistry.com/ACC-Shale-Report**

15 Annual energy outlook 2010 with projections to 2035. (2010, April) Energy Information Administration (EIA), AEO2010. p70. Retrieved from **http://www.eia.gov/oiaf/archive/aeo10/pdf/0383(2010).pdf**

16 Shale gas and new petrochemicals investment: Benefits for the economy, jobs and U.S. manufacturing. op. cit.

17   Millions of jobs from natural gas – and more possible!, Focus on jobs. NaturalGas.org. Retrieved from **http://www.naturalgas.org/jobs/jobs-home.asp**

18 Hampton, S. (2012, February 29) Blackstone invests in Cheniere Energy partners' LNG export plant. *Bizmology*, Business viewpoints from the editors at Hoover's. Retrieved from **http://bizmology.hoovers.com/2012/02/29/blackstone-invests-in-cheniere-energy-partners-lng-export-plant/**

19 Wingfield, B. (2012, April 17) Cheniere wins approval for biggest U.S. Gas-Export Terminal, Bloomberg. Retrieved from **http://www.businessweek.com/news/2012-04-16/cheniere-wins-u-dot-s-dot-approval-for-natural-gas-export-terminal**

CHAPTER VII
FRACTURING PROS AND CONS

Despite the rosy picture portrayed by extracting natural gas and petroleum from shale deposits, there are problems that must be confronted. The pressurized infusion of water and chemicals called hydrofracking pits environmental concerns against drilling interests.

There is a video that has gone viral on the Internet. It shows water coming from the kitchen sink catching fire because of the high amount of natural gas in the water supply.[1]

Affirming this is a documentary entitled *Gasland* that has won several awards about the consequences of deregulation in the drilling of natural gas.[2] It documents illness, hair loss, severe air pollution, and contamination of water wells, streams and rivers, resulting from "a new method of drilling developed by Halliburton…" There has been a pushback from the drilling industry seeking to discredit the film and its author, Josh Fox. Fox has a web site in which he counters arguments against him.[3]

New York has instituted a moratorium.[4] And Vermont has banned hydrofracking under all circumstances due to safety concerns.[5]

Water composes about 98% of what is pumped into the well, but the other 2% is made up of toxins and carcinogens which are dangerous to all life, humans, animals and plants. Companies have been reluctant to inform the public and the government what their mixtures are due to alleged issues of maintaining competitive advantages. If they disclose all of the chemicals that are shoved down the wells, then their competitors will know their secret ingredients. This rationale is based on the allegation that fellow gas and chemical companies aren't aware of what each other are doing which is a questionable assumption.

FRACKING NOT NEW

Old fashion fracking has been going on for decades but recent advances of horizontal drilling in massive new fields since the late 1980's have resulted in significant increases in extricating oil.[6] The fraternity of oil companies and their suppliers and servicing agents all know each other and it is doubtful that by now the processes they are keeping from each other are really secret. The type and number of chemicals that constitute 2% of the fracking solution is probably well known but the timing and the portion of each chemical concoction may be another matter.

"Through the Energy Policy Act of 2005 the US companies drilling for oil and gas were given an exemption from the Safe Water Drinking Act".[7] This exemption has greatly complicated the ability to insure that the fracking process becomes safe for the environment and for people. One negative consequence is that nondisclosure of the chemicals used in hydrofracking interferes with making the correct diagnosis and treatment of a patient who has been poisoned from hydrofracking residue.

The Fracturing Responsibility and Awareness of Chemicals Act was introduced in both houses of Congress in 2011 (H.R. 1084 and S. 587). It seeks to repeal the exemption for hydraulic fracturing in the Safe Drinking Water Act. It would require the energy industry to disclose the chemicals it mixes with the water and sand it pumps underground in the hydraulic fracturing process.[8, 9]

There is another condition that must be considered as well. When the water laced with toxins is pressurized deep underground, it dislodges ancient chemicals as well as the oil and natural gas. Everything is brought to the surface especially during the initial return of the fluids. The water now must be carefully disposed of for a longer period of time or cleansed of its dangerous elements, which increases the cost of the process.

FRACKING PROBLEMS

There have been incidents involving the following problems in disposing of the contaminated water.

- There is improper handling at the surface, which results in spillage onto the ground.[10]

- There is improper disposal into rivers or streams which damages the environment for animals, plants and people.[11]

- The carcinogenic substances are leaked due to improper sealing of the lining of the wells with weak cement or joints or via accidents. Toxins are also mixed into the original water concoction that is then pressurized to release the natural gas in shale deposits.[12]

- The water supply is contaminated through leakage at various steps of hydrofracking.[13, 14]

- Human made earthquakes have occurred or at least suspected to be caused by the injection of the fracking waste water into disposal wells.[15]

- Air pollution near natural gas wells is unhealthy for people, according to at least one scientific study and many reports from nearby residents.[16]

In general, oil and natural gas deposits are much further underground than water reservoirs which means that there should be a safe separation between the two deposits.

There are certainly many shale formations in rural or distant locations, which are away from people where hydrofracking can be securely performed. Of course, it is still important to properly handle and treat the contaminated water, which is an expensive but necessary process.

The above problems aren't minor in nature and must be confronted and resolved. All six of the examples above are due to human error, which can be solved by responsible companies..

MOST HYDROFRACKING SAFE

Ground water is near the surface and shale oil deposits are frequently thousands of feet underground. Extra care must be taken in bringing hydrofracking fluids to the surface. There have been many thousands of hydrofracking operations conducted without incident.

There is another problem due to the large amounts of water that are used in hydrofracking in territories that lack adequate aquatic resources. Texas is a good example of how water supply affects the implementation of hydrofracking. Some areas have a better supply of fresh water from rain and aquifers like in east Texas. However, other areas are suffering from harsh drought conditions such as in west and semi-arid conditions in central Texas.

Multiple pressures are exerting themselves since the state is experiencing a boom in drilling activity for natural gas and oil in such areas like the Barnett Shale and Eagle Rock shale plays. It takes millions of gallons of water to produce one hydrofracking job and thousands of these types of operations will be necessary in order to retrieve the natural gas and oil.

INCREASING DEMANDS

There have also been estimates that the population will double in the state by 2060 which will increase demands on existing water supplies. Conflicts are already occurring in how to allocate water and are projected to become more contentious in the future.

One underground aquifer in the Texas Panhandle (North Texas) is already exhibiting strain from overuse. This territory is serviced by the all important Ogallala aquifer which covers a wide geographical area extending into portions of eight states. This valuable water resource is being drained beyond the rate of replenishment in the Panhandle which creates a dilemma in how the resource should be distributed.[17]

DROUGHT CONDITIONS

In addition, the drought conditions in West Texas are even more severe and create greater clashes over water usage. Farmers, businesses and residents are suffering from the drought and want to either preserve the underground aquifer for the future or want to use it for their own purposes.[18]

If the drought persists, there won't be sufficient water for everyone. One approach in solving the water scarcity has been to construct massive projects to increase the number of reservoirs and to transport water via aqueducts. For instance 195 lakes in Texas were man-made constructions to satisfy the demands for storing water.[19]

Given the historical acceptance of large water projects, the state government recently proposed a massive effort to store, desalinate and distribute water at an estimated cost of $56 billion. The citizenry balked at the incredibly high price tag by voting down the proposal.[20] These actions place Texas in a major quandary because it is faced with having to simultaneously confront increasing demands under drought conditions.

The state must face up to population expansion, immense agricultural needs and now the promotion of hydrofracking by the oil and gas industry. Questions are being asked as to the suitability of using millions of gallons of water for every hydrofracking operation. All sides want to use water from the underground aquifer, which is claimed to be plentiful. On the other hand, if the drought persists there may not be any replacement water down the road.

NEW EPA STANDARDS

New EPA standards were issued this year (2012) on how to handle potential release of surface water and gas emissions at the well head.[21] The Fact Sheet is easily accessible and explains that responsible measures are mandated that reduce the number of harmful emissions or volatile organic compounds (VOC) including methane.[22]

Many companies already have implemented these practices. Other companies have yet to comply. To comply, it will be necessary to install equipment that will separate the gas and liquid substances and capture them for potential future sale. This "green completion" as it is labeled in the industry will"yield a 95 percent reduction in VOC emitted from more than 11,000 new hydraulically fractured gas wells each year."[23] Flaring will not be allowed at the well site.

There is an alternative to hydrofracking for many formations that consists in using a mixture of liquefied petroleum gas gel (LPG) in combination with other common chemicals.[24, 25] Sand is added to the final product which is then injected into the well for fracking purposes.

Gasfrac, a Canadian firm, which has exclusive rights to the process, has some 10 rigs using LPG in Canada and the United States.[26] The procedure avoids the use of water which automatically creates huge advantages over hydrofracking because all of the problems in disposing of contaminated fluids are avoided.[27, 28] In addition, a closed loop system eliminates ground exposure of all substances and the LPG doesn't disturb ancient toxic chemicals deep underground.[29]

The problems in the acceptance of LPG are that the process is more expensive; it has more safety concerns; and it is struggling against entrenched and larger competition. Although more than 1000 applications of LPG have occurred, the company is considered new to the field. There may be less pressure to adopt a more environmentally cleaner procedure since hydrofracking is exempted from the Safe Drinking Water Act of 2005.

END NOTES:
A companion web site has been established to go along with this report so that readers may view pictures, graphs and maps in color and link direct to referenced web sites.

The URL to this "Master" link is:

http://natgas-rpt.com

1 Cannon, S. (2011, December 10) Light your water on fire from gas drilling, fracking (video) *You Tube.* Retrieved at **http://www.youtube.com/watch?v=4LBjSXWQRV8**

2 GasLand (2010). IMDb, documentary, January 17, 2011. Retrieved from **http://www.imdb.com/title/tt1558250/**

3 Gasland (2010) Retrieved from **http://gaslandthemovie.com/**

4 Kastenbaum, S. (2012, May 2 updated). Fracking in New York: Risk vs. Reward, *CNN U.S.* Retrieved from **http://www.cnn.com/2012/03/09/us/new-york-fracking/index.html**

5 Associated Press (2012, May 18 updates). Vermont fracking ban: Green mountain state is first in U.S. to restrict gas drilling technique. *Huffington Post, The internet newspaper: News blogs video community.* Retrieved from **http://www.huffingtonpost.com/ 2012/05/17/vermont-fracking-ban-first_n_1522098.html**

6 Mark (2012, February 22) A short history of horizontal drilling for oil and natural gas, starting in the 1920's, and ending with Team Resources in 2011, *Oil and Gas News, Opinion and Analysis.* Retrieved from **http://www.oilgascenter.com/a-short-history-of-horizontal-drilling-for-oil-and-natural-gas-starting-in-the-1920s-and-ending-with-team-resources-in-2011/**

7 Water: Hydraulic fracturing: Regulation of hydraulic fracturing under the Safe Drinking Water Act. (2012, May 4). Environmental Protection Agency Retrieved from **http://water.epa.gov/ type /groundwater/uic/class2/hydraulicfracturing/wells_hydroreg.cfm**

8 You asked Congress for help on fracking, and they heard You! (2011, May 17). Alliance for Natural Health. Retrieved from **http://www.anh-usa.org/you-asked-congress-for-help-on-fracking-and-they-heard-you/**

9 H.R. 1084 Responsibility and awareness of chemicals act of 2011, To repeal the exemption for hydraulic fracturing in the Safe Drinking Water Act and for other purposes. Open Congress for the 112th Congress of the United States. Retrieved from **http://www.opencongress.org/bill/112-h1084/news_blogs**

10 Walsh, B. (2011, April 20). More problems with fracking-and some solutions, *Time, Science & Space.* Retrieved from **http://science.time.com/2011/04/20/more-problems-with-fracking-and-some-solutions/**

11 Urbina, I. (2011, February 26) Regulation lax as gas wells tainted water hits rivers, *The New York Times.* Retrieved from **http://www.nytimes.com/2011/02/27/us/27gas.html?pagewanted=all**

12 Faulty Cementing (2011, May 17) Save Colorado from fracking, A coalition of citizens who love Colorado. Retrieved from **http://www.savecoloradofromfracking.org/ whatgoeswrong/cementing.html**

13 By the Editors (2012, February 22) Maximize promise of fracking by solving safety problems: View. *Bloomberg, Opinion.* Retrieved from **http://www.bloomberg.com/news/2012-02-23/maximize-promise-of-fracking-by-solving-safety-problems-view.html**

14 Osborn, S. G., Vengosh, A., Warner, N.R. & Jackson, R.B. (2011, April 14) Methane contamination of drinking water accompanying gas-well drilling and hydraulic fracturing. Proceedings of the National Academy of Sciences of the United States of America (PNAS). Retrieved from
http://www.pnas.org/content/early/2011/05/02/1100682108

15 Efstathiou, J. Jr. (2012, April 20) Fracking-linked earthquakes spurring state regulations. *Bloomberg.* Retrieved from
http://www.bloomberg.com/news/2012-02-23/maximize-promise-of-fracking-by-solving-safety-problems-view.html

16 Tollefson, J. (2012, February 7) Air sampling reveals high methane emissions from natural gas field, Methane leaks during production may offset climate benefits of natural gas, *Scientific American.* Retrieved from **http://www.scientificamerican.com/article.cfm?id=air-sampling-reveals-high-meth**

17 Galbraith, K. (2010, June 17) How bad is the Ogalalla's aquifer's decline in Texas?, *The Texas Tribune.* Retrieved from
http://www.texastribune.org/texas-environmental-news/water-supply/how-bad-is-the-ogallala-aquifers-decline-in-texas/

18 Infrastructure: Water, Texas in focus: A statewide view of opportunities. Window on State Government, State Comptroller of Public Accounts, Susan Combs. Retrieved from
http://www.window.state.tx.us/specialrpt/tif/water.html

19 Infrastructure: Water, Texas in focus: A statewide view of opportunities. Ibid.

20 Tresaugue, M (2011, November 12) Texas water supply for the future is uncertain, *Houston Chronicle.* Retrieved from
http://www.chron.com/news/houston-texas/article/Where-s-tomorrow-s-water-2266277.php#page-3

21 Regulatory Actions (2012, April 20) Oil and natural gas air pollution standards, EPA issues final air rules for the oil and gas industry. Environmental Protection Agency (EPA). Retrieved from
http://epa.gov/airquality/oilandgas/actions.html

22 Fact Sheet (2012, April 17) Overview of final amendments to air regulations for the oil and natural gas industry. Environmental Protection Agency (EPA). Retrieved from
http://epa.gov/airquality/oilandgas/actions.html

23 Fact Sheet (2012, April 17). Ibid.

24   Chameides, B. (2012, May 15) Another 'game-changer' for natural gas, Not fracking with millions of gallons of water this time around— waterless fracking. *Scientific American.* Retrieved from **https://www.scientificamerican.com/article.cfm?id=another-game-changer-for-natural-ga**

25 Schoonover, N. GasFrac gelling video, This is a visual demonstration of the process of converting Liquefied Petroleum Gas to a gel to carry sand to fractures. Retrieved from   **http://vimeo.com/25013243**

26 Numerous patents/ pending, Proven innovative LPG technology. GasFrac Energy Services Inc. Retrieved from **http://www.gasfrac.com/proven-proprietary-process.html**

27   Bachhav, K. (2012, April 4) Liquid propane – Upcoming eco-friendly fracturing technique, LPG Frack. Retrieved from **http://world-petroleum-techtrends.blogspot.com/2012/04/normal-0-false-false-false-en-in-x-none.html**

 28   Bino, A. & Nearing, B. (2011, November 6) New waterless fracking method avoids pollution problems, but drillers slow to embrace it. *Inside Climate News.* (Produced in partnership with the Albany (New York) Times-Union. Retrieved at **http://insideclimatenews.org/news/20111104/gasfrac-propane-natural-gas-drilling-hydraulic-fracturing-fracking-drinking-water-marcellus-shale-new-york?page=show**

29 Safe and efficient, A completely closed system with automated remote operations. Gasfrac, LPG Technology, Safer energy solutions. Retrieved from **http://www.gasfrac.com/safer-energy-solutions.html**

CHAPTER VIII
CONVERTING TO NATURAL GAS

There are five types of gas fuels that are being utilized in transportation, heating and generation of electricity in America. They are natural gas, methane, propane, butane and hydrogen. Natural gas and to a lesser extent methane is what this report is about.

Methane is natural gas without the impurities. Up to 20 percent of raw natural gas (gas as it is produced from wells) consists of impurities that must be refined out before it can be used as fuel. Propane and butane are byproducts of the refinement of natural gas.

Approximately 1,800 landfills produce and release methane as part of the normal process of waste decomposition. Some public municipalities have constructed extraction processes to put the natural gas to good use.

For instance, Waste Management, a major refuse/trash company, "had 119 landfill-gas-to-energy projects producing 540 megawatts of power, the equivalent of powering approximately 400,000 homes" by the end of 2009.[1]

Their largest venture was a cooperative effort with Linde North America that resulted in constructing the world's largest facility of its kind for transforming landfill waste into 13,000 gallons of LNG per day. This was sufficient amount of fuel to power 300 LNG trucks daily.[2]

Other agencies have ignored this opportunity or even flared off the excess methane which is damaging to the earth's atmosphere. Methane is about 20 times more detrimental to the earth's atmosphere in comparison to an equivalent volume of carbon dioxide from petroleum based fuels..

Fortunately, methane constitutes approximately 10 percent of the particulates in comparison to nitrogen oxides from oil products that make up the vast majority of the greenhouse effect (90 percent). This characteristic means that methane damages the ozone layer and creates a greenhouse effect that increases the earth's temperature. Therefore, preventing leakage of methane into the atmosphere from all controllable causes such as from landfills and gas and oil rigs is important to the health the planet.[3, 4]

CONVERSION

Converting vehicles to operate on natural gas is different depending on the sizes and types of vehicles. The industry has developed a classification system ranging from one through eight which is based on a vehicles gross weight. Thus the classifications for all vehicles and their weight limits consist of the following:

Classified as light duty are Classes 1, 2 and 3
        Class 1, up to 6,000 lbs.;
        Class 2. 6,000 to 10,000 lbs.;
        Class 3. 10,000 to 14,000 lbs.

Classified as medium duty are Classes 4, 5 and 6.
        Class 4, 14,000 to 16,000 lbs.;
        Class 5, 16,000 to 19,500 lbs.;
        Class 6, 19,500 to 26,000 lbs.

The first four classes are light passenger cars, pickup trucks and SUV's manufactured by the big three auto-makers. Classes 5 and 6 are medium duty trucks, utility vans and transit vehicles, etc..

Classes 7 and 8 are heavy duty trucks manufactured by the big four truck manufacturers. They are Daimler Volvo, Paccar and Navistar.

Class 7 weight is 26,000 to 33,000 lbs., and
Class 8 weight is anything above 33,000 lbs.

Class 8 trucks include tractors and dump trucks. Volvo is based in Greensboro, North Carolina and manufactures class eight vehicles. Paccar, which is based in Portland, Oregon, manufactures medium and heavy duty trucks under the brands Kenworth, Peterbilt and DAF

FIVE KINDS OF ENGINES

There are five kinds of engines that have to be considered in converting to natural gas. They are bi-fuel or bivalent, dual fuel, spark ignited (mono fuel or monovalent), tri fuel and high pressure direct injection (HPDI)

Westport Innovative, a Canadian engine technology firm, has developed and patented a new high pressure direct injector (HPDI.)[5] It calls it "A revolutionary patented injector with a dual-concentric needle design". It allows for small quantities of diesel fuel and large quantities of natural gas to be delivered at high pressure to the combustion chamber. The natural gas is injected at the end of the compression stroke.[6] Natural gas requires a higher ignition temperature than diesel. To assist with ignition, a small amount of diesel fuel is injected into the engine cylinder followed by the main natural gas fuel injection. The diesel acts as a pilot, rapidly igniting the hot combustion products, and thus the natural gas. HPDI replaces approximately 95% of the diesel fuel (by energy) with natural gas.[7]

Westport HPDI technology may be the only engine technology that maximizes natural gas use while maintaining equal horsepower, torque and efficiency characteristics of a diesel-fueled engine. In the past, natural gas engines produced less power because diesel fuel ignites and burns at a lower temperature.

MAJOR ENGINE COMPANIES

Westport has partnered with major engine companies such as Cummins, an American company, Daimler Trucks International and Weichai, a Chinese engine manufacturer.[8-10] In addition, Navistar and Paccar Cos (Peterbilt and Kenworth) have joined in developing natural gas engine technology.[11-13]

There is a partnership with Canadian National Railroad for creating locomotives that will operate on natural gas and two such engines are currently being tested.[14] They have also established business contracts with companies that manufacture busses, refuse/trucks and transit vehicles.

Pickup trucks by Ford were introduced in 2012 and future projects with GM are planned.[15, 16] Development of large construction equipment powered by natural gas in cooperation with Caterpillar is underway as well.[17]

Power Solutions International, an American company, has specialized in designing and manufacturing natural gas engines for industrial uses but has now expanded its product line to provide powertrains for Class 4, 5, 6 and 7 vehicles. Their new market encompasses light to medium duty vehicles such as delivery vans, refuse/garbage vehicles, and school/transit buses. Key benefits to these engines include durability, cost, fuel-flexibility, power, efficiency, and emissions compliance.[18]

In Europe Volvo has teamed up with Clean Power to produce dual fueled trucks that operate on either natural gas and/or diesel The savings aren't as great with the single fueled natural gas engines but there is added security for the truck to use diesel since the infrastructure for natural gas is inadequate.[19, 20]

In addition, companies such as Fuel System Solutions are converting cars to natural gas engines. They have negotiated a contract with GM to manufacture pickup trucks to compete with Ford.[21, 22]

The Environmental Protection Agency (EPA) during the past year has updated their regulations and costs for certifying natural gas engines that can replace an entire class of petroleum based engines. For example, a manufacturer submits a natural gas engine which it has designed to replace all of Toyota's Camry and Avalon models to the EPA for certification.

The governmental agency completes a series of tests to determine that the natural gas engines are suitable for powering the stated vehicle models, and that these engines are no more polluting than the original petroleum based engines. All power trains must meet the same EPA standards for pollution controls.

The above processing occurs once and is valid in all 50 states. There is increased flexibility in approval and the cost can be spread out over however many vehicles the manufacturer can convert to natural gas. The certification remains the same for any number of vehicles so the per unit expense declines as the volume increases.

The ability for companies such as Fuel Systems Solutions to convert engines to natural gas should be easier in the future given the new EPA regulations.

Janet Cohen of the EPA's Office of Transportation and Air Quality, has expressed concern that the new regulations are not fully understood. In an email to Western Research Institute, she said, "There is a great deal of misinformation circulating about EPA's requirements for alternative fuel conversions."

She said the "EPA does not require a permit or charge a fee to convert individual vehicles and engines. Conversion manufacturers do need to show EPA that their technology will not increase pollution. The demonstration applies to a group of vehicles and engines with similar characteristics. In some cases, this demonstration involves testing and/or payment of a fee.

"Under the new regulations, if the fuel converted vehicle is new (less than two years old), a one-time EPA fee is charged ranging from under $1000 to several thousand dollars. The fee is paid only upon initial certification. Once a conversion technology is certified, the manufacturer may continue to sell the kit in future years without re-certifying or paying new fees. If the fuel converted vehicle is more than two years old, the manufacturer may choose an alternative compliance pathway that does not require any fee to be paid."[23]

FUEL TANKS

There are issues concerning the capacity and safety of fuel tanks for natural gas vehicles. Compressed natural gas (CNG) takes more volume to obtain equivalent amount of power in comparison to petroleum based fuels.

The first solution for shorter distances required for pickup trucks, vans and cars consists of using compressed natural gas (CNG). It becomes necessary to engineer larger storage tanks that are thick enough to withstand the increased pressure. It would also be an improvement if the tanks were made out of lighter, higher tech materials which meet the same demanding safety requirements since less weight would be an advantage.

Much longer traveling distances are necessary for heavy duty transportation vehicles that cover hundreds of miles per day. In this instance it is possible to construct a much larger set of compressed natural gas tanks that cover the entire back end of the cab. These CNG tanks have a new configuration in comparison to where the previous petroleum gas tanks were placed.[24]

LIQUID NATURAL GAS

The second solution for long distance travel involves using liquefied natural gas (LNG) which is much denser and heavier in comparison to CNG. This means that greater volume of the fuel can be stored in a tank which results in an extended driving range. These fuel tanks must be capable of carrying heavier weights due to the liquefied fuel and also must possess hydraulic mechanisms to push selective amounts of fuel out of the tank at the precise time for use in the engine. The demonstration that the author was shown indicated that a small amount of LNG from the storage tank is injected into a separate holding container and a brief time later the fuel reaches a gaseous state through a natural heating process. The natural gas is now ready to enter the engine for combustion purposes.

The ability to keep LNG cold for long periods of time, days or weeks, is solved in the following manner: LNG becomes frozen at minus 259 degrees Fahrenheit (-161 degrees Celsius), and it remains liquid for a few weeks before most of it returns to its original natural gas state. However, if LNG continues to be pumped into the tank frequently it remains in a liquid condition.[25]

Therefore the correct procedure involves the following steps. Whenever the truck is driven, LNG fuel must be added at the end of the day to refill all of its tanks and to maintain the gas in a liquid state. If the vehicle remains idle, then on a periodic basis the small amount of LNG that has returned to a gaseous condition should be vented back into the main storage container. An equivalent amount of LNG should be added to the truck's tank in order to maintain full capacity and to sustain the necessary frozen temperature.

All natural gas vehicles and trucks use thicker and stronger tanks as a safety measure. Natural gas is actually safer than petroleum products in most situations since it ignites at a higher and narrower temperature range and it evaporates when exposed to the air.[26] Natural gas is so safe under normal conditions that many households use it to cook with and to heat both their water and homes. Gasoline is too dangerous to use for such purposes.

Another safety issue has to do with spills or leaks from fuel tanks. When gasoline and diesel fuel spills occur, the substance always remains on the ground and pools form that are hazardous until cleaned up. Natural gas which is lighter than air dissipates upward into the atmosphere if there is a leak or spill in open space.

The case is made in Popular Mechanics magazine that if gasoline was newly invented today, it wouldn't be used as a transportation fuel because it would be considered too dangerous.[27]

EXPLOSIONS A CONCERN

Explosions from natural gas are a concern, so precautions are necessary. Most of the accidents have occurred in pipeline incidents where leaks of natural gas within a closed area result in an explosion when ignited by a spark or other means.

Cars and trucks powered by natural gas must adhere to the same safety standards as petroleum based vehicles and there is no increased risk of an explosion in an accident. Nationwide Insurance stated as early as 1992 that they considered natural gas vehicles safe and would insure them in the same manner and price level as gasoline or diesel powered vehicles.[28] To our knowledge, natural gas trucks are insured on the same basis as any other vehicle due to their safety characteristics. Natural gas storage tanks should be inspected at regular intervals for corrosion.[29]

METHANE

In regard to methane, there are environmental hazards to people who live close to landfills. If escaping methane is not captured, then their homes and health will be adversely affected. Regardless of whether or not the methane is used as a fuel source, there is a need to avoid contamination.

One obstacle to constructing elaborate systems to capture the methane is the low price of natural gas in America. At $2 equivalent gallon of gasoline energy, the municipalities may not be competitive in their pricing given the costs to build and implement the systems.

ESCAPING METHANE

The dilemma exists about what to do with the methane escaping from the landfills? Given the health and environmental hazards, there should be efforts to capture, purify and use the methane. The pay-back time may be extended given the economics but the safety and liability exposure to lawsuits will certainly make up for the expense. The landfills could become a long term secondary source for generating methane/natural gas if issues occur in the supply of natural gas reserves in the distant future. The landfills become a method to naturally manufacture natural gas which should be implemented.

One study in 1998 by the Department of Energy was summarized on the NGVA web site indicating that sufficient methane could be captured from landfills to provide 25 percent of the diesel fuel used by buses and freight trucks (1998 mileage).[30]

The logical approach to determining whether or not to convert a landfill into a methane capturing system depends on calculating the factors of safety and profitability. Those landfills that are the most hazardous to humans and environment should be given first priority and those sites which are the most economical or profitable are second choices. The calculations should be made on this matrix with corresponding values.

One example of this process of converting biomass or in this case manure from cows in the dairy industry originates from Fair Oaks Farms in Indiana where CNG stations will be completed for fueling the delivery trucks and other heavy duty vehicles.[31, 32]

The station is open to the public and is an example of public/private partnership. The EPA has studied the conversion process in purifying the escaping methane from cows and landfills and has indicated that it is a safe and feasible goal. It is expected that similar projects will be replicated elsewhere throughout the country

Additional progress has been achieved in converting natural gas from biomass and cleanup from landfills by two cooperating companies under the guidance from the public sector. Acrion Technologies has a proprietary CO_2 Wash system that purifies the captured methane gas and FirmGreen Inc. has a process of converting the methane into CNG or LNG fuel. Successful joint projects have been completed in Columbus, Ohio and internationally in Brazil.[33, 34]

The Argonne's Transportation Technology Research and Development Center which is part of the Department of Energy provided grant money and support to assist in creating the methane purification process. This is another example of successful teamwork between the public/private sectors to generate new technology which is then applied on a market basis.[35]

There are other sources of biomass that generate energy and fuel which consists of burning material such as wood chips, grass and other plant products. This process could be efficient if the biomass is being generated as an inherent part of a manufacturing process. For example, a saw mill or furniture manufacturer could effectively use wood chips which are always created during their operation for generating energy.

Genetic engineering is another possibility to produce fuels from sources like plants and algae. Sugar cane is also utilized for ethanol production a substance that extends the mileage of standard gasoline. Discussion of the complexities of these sources of biomass is not the subject of this book and requires a separate analysis.

In summary, many exciting developments using natural gas as a transportation fuel are occurring, but a survey by PLS Logistics indicated that there is a slow adoption by industry. A majority of respondents stated that the lack of refueling infrastructure was the primary obstacle to their purchase of heavy duty natural gas trucks.[36]

END NOTES:
A companion web site has been established to go along with this report so that readers may view pictures, graphs and maps in color and link direct to referenced web sites.

The URL to this "Master" link is:

http://natgas-rpt.com

1 Waste Management named EPA landfill methane outreach program industry partner of the year. (2012, January 18) Waste Management Inc. *About Us, Press Room*, 2012.   Retrieved from **http://www.wm.com/about/press-room/2012/20120118_LMOP.jsp**

2 Linde and Waste Management receive California governor's award for sustainable facility. (2010, November 16) Waste Management press release, p.1. Retrieved from **http://www. wm.com/about/press-room/pdfs/20101117_WM_Linde_ Receive_Sustainability_Award_From_California_Governor.pdf**

3 Koerner, B. (2007, November 27) The other greenhouse gases. *Slate*. Retrieved **from http://www.slate.com/articles/health_and_ science/the_green_lantern/2007/11/the_other_greenhouse_ gases.html**

4 How does global warming occur? Global Warming, continued. Ecosystems, Hazards to Our World. Retrieved from **http://peer.tamu. edu/curriculum_modules/ecosystems/hazards/global_warming2.htm**

5 Westport, the global leader in natural gas engines. Retrieved from **http://www.westport.com**

6 Westport HPDI Technology. Westport Innovative. Retrieved from **http://www.westport.com/is/core-technologies/hpdi**

7 Westport HPDI Technology, ibid.

8 Cummins Westport, The natural choice. Retrieved from **http://www.cumminswestport.com/**

9 Natural gas: CNG and LNG, Daimler Trucks North America. Retrieved from **http://www.daimler-trucksnorthamerica.com/news /featurearticle/alternativefuels/natural-gas.aspx**

10 China's first engine with Westport HPDI technology unveiled during China National People's Congress. (2012, March 14) *NGV Global News*. Retrieved from **http://www.ngvglobal.com/chinas-first-engine-with-westport-hpdi-technology-unveiled-during-china-national-peoples-congress-0314**

11 Partnership to promote natural gas. Navistar Retrieved from. **http://www.internationaltrucks.com/trucks/naturalgas**

12 Design, Natural gas, (2011) Peterbilt, A PACCAR Co. Retrieved from **http://www.peterbilt.com/eco/Design-LNG.htm**

13 Kenworth to add heavy duty Cummins Westport ISX12 G to complete line of natural gas engines, (2012, March 21) Kenworth news release, A PACCCAR Co. Retrieved from **http://www.kenworth.com/news/news-releases/2012/march/kenworth-to-add-heavy-duty-cummins-westport-isx12-g-to-complete-line-of-natural-gas-engines.aspx**

14 CN tests natural gas/diesel fuel powered locomotives between Edmonton and Fort Mc Murray, Alta. (2012, September 27) CN (Canadian National Railway Co.). Retrieved from **http://www.cn.ca/en/media-news-cn-tests-natural-gas-locomotives-20120927.htm#.UGW4tpCYkik.twitter**

15 Westport Wing™ Power Systems for Ford F-250 and 350 series rolls out at NTEA Show in Indianapolis. (2012, March 6) Westport Innovative. Retrieved from http://www.westport.com/news/2012/westport-wing-power-systems-for-ford-f-250-and-350-series-rolls-out-at-ntea-show-in-indianapolis

16 Westport announces second agreement with General Motors for advanced engineering development for light-duty vehicles. (2012, June 27) Westport Innovative. Retrieved from http://www.westport.com/news/2012/second-agreement-with-general-motors-for-advanced-engineering-development-for-light-duty-vehicles

17 Westport and Caterpillar announce agreement to develop natural gas technology for off-road equipment. (2012, June 5) Westport Innovative. Retrieved from http://www.westport.com/news/2012/westport-and-caterpillar-natural-gas-technology-agreement

18 Our Companies, Power Solutions, Inc. and Natural gas engines, Power Solutions International. Retrieved from http://powersint.com/our-companies/

19 A high performing, competitive vehicle – and a sustainable solution. VOLVO Trucks Global. Retrieved from http://www.volvotrucks.com/trucks/global/en-gb/trucks/new-trucks/Pages/volvo-fm-methanediesel.aspx

20 Clean Air Power is the leading company moving towards using natural gas as a fuel for heavy goods vehicles. Clean Air Power. Retrieved from http://www.cleanairpower.com/

21 GM Offering bi-fuel pickups, Heavy duty trucks will run on natural gas and gasoline (2012, March 5), IMPCO Automotive, Power Solutions Inc. Press release. Retrieved from http://www.impcoautomotive.com/index.php?pagename=pr-GM%20bi%20fuel%20truck

22 Technology for a cleaner future – Today! Fuel Systems Solutions, 2010. Retrieved from http://www.fuelsystemssolutions.com/

23 The Clean Alternative Fuel Vehicle and Engine Conversions Final Rule. (2011, April 8). New regulations significantly increased flexibility and streamlined compliance processes for alternative fuel converters. Environmental Protection Agency (EPA). Links to full text of the rule, as well as summaries and additional information for converters can be found on the EPA alternative fuels conversions web site. Retrieved from http://www.epa.gov/otaq/consumer/fuels/altfuels/altfuels.htm The cost analysis is found in section 7 and 8 starting p. 19853.

24 Piellisch, (2012, March 26) Freightliner Cascadia 113 with ISX12. Show Times, Clean Fuels & Vehicle. Retrieved from **http://showtimesdaily.com/fleetsfuels/freightliner-cascadia -113-with-isx12-g** (Diagrams of fuel tanks can be viewed by double clicking on first photo for enlarged picture of fuel tanks attached to back of cab.)

25 Frequently asked questions about LNG. CA.GOV, The California Energy Commission. Retrieved from **http://www.energy.ca.gov/lng/faq.html**

26 How safe are natural gas vehicles? (2010, September 17 revision) Technology Committee Bulletin, Clean Vehicle Education Foundation, September 28, 1999. Retrieved from **http://www.cleanvehicle.org/committee/technical/PDFs/Web-TC-TechBul2-Safety.pdf**

27 Allen, M. (2008, September 15) Pickens overlooks existing natural gas cars in energy plan: Reality check. *Popular Mechanics*. Retrieved from **http://www.popularmechanics.com/cars/alternative-fuel/4282954**

28 How safe are natural gas vehicles? Ibid.

29 Green, C. Natural gas vehicle safety. *eHow, Discover the expert in you*. Retrieved from **http://www.ehow.com/about_5455641_natural-gas-vehicle-safety.html**

30 NGV's and biomethane. Natural Gas Vehicles for America, (NGVAmerica). Retrieved from **http://www.ngvamerica.org/about_ngv/ngv_biomethane.html**

31 What is black and white but green all over? (2011, June 14) Oaks Farms, Indiana. Retrieved from **http://www.fofarms.com/PDF/GREENFOF.pdf**

32 Sober, D. L. (2009, February 24-25) Biomethane from dairy waste conversion: Guidance for introduction to natural gas networks, 2009 AgSTar National Conference. Retrieved from **http://www.epa.gov/agstar/documents/conf09/agstar_saber_presentation.pdf**

33 Acrion Technologies, CO_2 Wash Design and Development Services. Retrieved from **http://www.acrion.com/**

34 FirmGreen Inc. Retrieved from **http://www.firmgreen.com/**

35 Werpy, M. R., Santini, D., and Mintz, M. (2010, August) Natural Gas Vehicles: Status, barriers and opportunities. Center for Transportation Research, Energy Systems Division, Argonne National Laboratory, p. 24-27. Retrieved from
http://www.transportation.anl.gov/pdfs/AF/645.PDF

36 Use of LNG-Powered vehicles for industrial freight: Carrier survey results show high interest, but slow adoption. PLS Logistics Services, 2012. Retrieved from
http://www.plslogistics.com/whitepapers/PLS_LNG_Study.pdf

CHAPTER IX
NATURAL GAS HIGHWAY

Presently, there are too few compressed natural gas fueling stations across the U.S. to make mass production of CNG vehicles feasible. However, this is about to change.

The number of natural gas fueling stations is expected to increase by as much as 40 percent in just a few years. A number of entities have announced plans to combine forces to build and operate large networks of LNG stations.

At the present time there is an estimated 1100 LNG stations across the country. It is difficult to nail down statistics of the exact number as they keep growing and reporting entities do not agree with each other.

Gasoline and diesel are measured in gallons. Natural gas is measured in cubic feet. The British thermal unit (Btu) is used to measure heat energy. It is the most commonly used unit for comparing fuels in this country.

A gallon of gasoline equivalent (GGE) equals 124,238 Btu and a gallon of diesel equals 138,690 Btu. A cubic foot of natural gas equals 1023 Btu. Thus it takes a little more than 121 cubic feet of natural gas to supply the same energy as a gallon of gasoline and a little more than 135 cubic feet to supply the same energy as a gallon of diesel.

This is the natural gas equivalent of a gallon of gasoline and of diesel that is derived from figures on the Energy Information Administration web site (October, 2012).[1]

Manufacturers won't build large numbers of natural gas vehicles unless there are enough buyers for them. There won't be enough buyers as long as there is a lack of refueling stations.

PROGRESS MADE

There is progress being made. It is now possible to go across the country on CNG. The web site of CNG Now (cngnow.com) has an interactive map that supposedly lists the locations of all CNG refueling stations across the country and the cost per gallon equivalent of diesel. Interestingly the price varies from $1.11 to $1.99 around Oklahoma City to from $1.91 to $2.90 in the Los Angeles CA area.[2] There are also other web sites that list locations of CNG stations. [3]

Prominent among companies announcing plans to build new stations is Clean Energy, headquartered in Seal Beach, California. In January of 2012, it announced that it was building 150 LNG stations that would span the country by the end of 2013. It said it "has identified 98 locations and anticipates having 70 stations open by the end of 2012 in 33 states. [4]

"Many of the fueling stations will be located at Pilot Flying J-Travel Centers…" It added, "Pilot-Flying J is the nation's largest truck-stop operator with more than 550 retail properties in 47 states."[5]

A major stockholder in Clean Energy is T. Boone Pickens, a multi-billionaire Texas oilman. He is a leading figure in the movement to convert to natural gas as a fuel for transportation.

Clean Energy is listed on the NASDAQ Stock Exchange under the ticker symbol CLNE.

According to its web site, "Clean Energy is the largest provider of natural gas fuel for transportation in North America and a global leader in the expanding natural gas vehicle market. It has operations in CNG and LNG vehicle fueling construction and operation of CNG and LNG fueling stations, bio-methane production, vehicle conversion and compressor technology." LNG or liquefied natural gas becomes a heavy liquid at extreme freezing temperatures.[6]

Other companies have announced plans to combine forces to build and operate natural gas fueling stations as well. Shell Oil and Travel Centers of America (TA) will construct 200 LNG pumps at 100 different locations along the country's freeway system and in metropolitan areas starting in 2013. [7]

GE and Peake Fuel Solutions, a subsidiary of Chesapeake Energy, have developed a system that they call 'CNG in a Box'. This is a fueling station that contains a modular standardized natural gas compression system designed for vehicles. Their strategy is to initially place 250 of these pumps at numerous unspecified locations starting in the fall of 2012 and ending in 2015.[8]

Chesapeake Energy Corporation says it is the second largest producer of natural gas, a Top 15 producer of oil and natural gas liquids and the most active driller of new wells in the U.S. It is headquartered in Oklahoma City and is seeking to create a wider market for its natural gas resources by applying them as a transportation fuel on a mass basis.[9]

The number of natural gas stations that were built in 2011 is unknown at this time but given the announcements by major corporations like Shell Oil and Chesapeake Energy, there will be a significant increase in natural gas infrastructure during the next two years.

America's Independent Trucker's Association estimates that there are almost 5,000 truck stops in America.[10]   If you add up the number of locations of the partnerships discussed previously involving Shell Oil, Clean Energy and Chesapeake Energy etc., they plan on building natural gas pumps at a maximum of 500 different locations during the next 2 to 3 years. This amounts to 10% of the number of truck stops which is a major improvement but hardly substantial market penetration.

HOME FUELING OPTIONS

For private vehicle owners of passenger cars and pickups, there is the option of a home natural gas fueling system. One such is produced by BRC Fuelmaker, a division of Fuel Systems Solutions. It has forty-one dealerships in the U.S. which are listed on the company's web site.[11] Fuel Systems Solutions specializes in converting gasoline engines to operate on natural gas fuel.

Honda is currently the only manufacturer of passenger cars in the United States that are said to run on natural gas.

END NOTES:
A companion web site has been established to go along with this report so that readers may view pictures, graphs and maps in color and link direct to referenced web sites.

The URL to this "Master" link is:

http://natgas-rpt.com

1. The number of British Thermal Units assigned to a cubic foot of natural gas and a gallon of gasoline and diesel vary depending on temperature, pressure and other factors. Thus web sites are not in agreement. A similar, but different set of figures may be found at Gable, Christine and Scott Fuel energy comparisons: Gasoline gallon equivalents (GGE), *About.com, Hyrbid Cars & Alt Fuels*. Retrieved from **http://alternativefuels.about.com/od/resources/a/gge.htm**

2 Stations, Looking to fill up or just wondering where you would go if you had a CNG vehicle? *CNGnow.com.* Retrieved from **http://www.cngnow.com/stations/Pages/information.aspx**

3 Alternative fuel station locator. U.S. Department of Energy, Alternative fuels data center, locate stations. Retrieved from **http://www.afdc.energy.gov/locator/stations/**

4 Clean energy unveils backbone network for America's natural gas highway – 150 LNG truck fueling stations enabling goods movement coast-to-coast and border-to-border anticipated by end of 2013 –. (2012, January 12) Clean Energy Fuels. Retrieved from **http://www.cleanenergyfuels.com/2012/1-12-12.html**

5 Clean energy unveils backbone network for America's natural gas highway. Ibid.

6 Clean Energy annual report 2011. p 1. Retrieved from **http://www.cleanenergyfuels.com/annualreport/2011/atglance.html**

7 Cassidy, W.B. (2012, June 7) Shell, TA sign natural gas pact. *Journal of Commerce Online.* Retrieved from **http://www.joc.com/suppliers/shell-ta-sign-natural-gas-pact**

8 GE and Chesapeake Energy Corporation announce collaboration to speed adoption of natural gas as transportation fuel", (Accessed June 2012), Chesapeake Energy, March 7, 2012. Retrieved from **http://www.chk.com/News/Articles/Pages/1670077.aspx**

9. About Chesapeake Energy, Who we are, (Accessed May 2012), Chesapeake Energy. **http://www.chk.com/About/Pages/Default.aspx**

10 AITA, American Independent Truckers' Association, Inc. Retrieved from **http://www.aitaonline.com/TS/Locations.html**

11 BRC FuelMaker, Dealers USA, all. Retrieved from **http://www.brcfuelmaker.it/eng/casa/ricerca.asp?nazione=U.S.A.&regione=Tutte&provincia=Tutte**

CHAPTER X
FAILING GRADE FOR U.S.

A confluence of events including the discovery of expansive oil shale deposits coupled with new drilling techniques and the development of new natural gas truck engines provide the U.S. with the means of becoming oil independent. So why is the United States lagging so far behind other countries in converting to natural gas for transportation?

The lack of progress in utilizing natural gas for transportation is even more puzzling given the disproportionate amount of oil that the United States consumes which is 22.5% of global consumption.[1] America represents only 4.5% of the world's population.[2]

As of 2011, there were 13 countries that had more natural gas vehicles on the road than the United States. This despite the U.S. being fifth in the amount of natural gas reserves (behind Russia, Iran, Qatar and Saudi Arabia) and having the world's most advanced vehicular transportation system.

The top two countries were Pakistan and Iran which have some 20 times the number of natural gas vehicles on the road than America with Argentina, Brazil and India close behind. The United States ranks fifth in the number of refueling stations.[3]

The number of natural gas vehicles (NGVs) on the road today in the U.S. and the number of NGV fueling stations in the U.S. is difficult to come by. There are a number of web sites that list conflicting figures.[4] The figures used here are taken from the web pages of NGV Global which cautions that these numbers are only estimates.

NUMBER ESTIMATED

Our estimation is there are more than 123,000 NGVs and more than 1,100 fueling stations that are open to the public as of last year (2011). In addition, the numbers of natural gas vehicles that are classified in size as being light, medium and heavy duty are difficult to ascertain. The reported data isn't clear as to what classes of vehicles are counted and not counted.[5]

The global transformation to natural gas vehicles is quite a different story. There have been increases every year since 2000 with notable expansion of 20 percent in 2010 and 33 percent in 2011.[6] The following data lists countries ahead of the United States as of 2011. The data lists the name of the country, the total number of vehicles on the road in that country, the number of fueling stations in that country, the percentage change from 2010 and the number of vehicles on the road in that country 2010.

Country	Vehicles	Stations	Percentage Change
Iran	2,859,386	1,574	46% incr.
	from 1,954,925		
Pakistan	2,850,500	3,285	4% incr.
	from 2,740,000		
Argentina	1,900,000	1,878	-0.1% decr
	from1,901,116		
Brazil	1,694,278	1,725	1.8% incr.
	from 1,664,847		
India	1,100,000	571	1.9% incr.
	from 1,080,000		

China	1,000,000 from 450,000	1,350	122% incr.
Italy	779.000 from 730,000	790	-1.4% decr.
Columbia	348,747 from 340,000	614	2.6% incr.
Thailand	300,581 from 218,459	426	37.6% incr.
Ukraine	390,000 from 200,000	285	95% incr.
Bangladesh	203,000 from193,521	546	5% incr.
Bolivia	157,426 from 140,400	156	12% incr.
Egypt	157,858 from 122,271	119	29% incr.
United States	123,000 from 112,000	1,000	9.8% incr.

One reason for the anemic growth in the manufacture and sale of natural gas vehicles in the United States is the disinterest of the federal government. There have been four versions of the natural gas act introduced in Congress in the last four years (2009-2012), in the Senate and in the House in the 111th and 112th Congress. None have made it out of committee.

There has however been a vote to attach the provisions of the Senate natural gas bill as an amendment to the Transportation Bill in March 2012. Although the motion had a majority of votes of 51 for and 47 against with 2 not voting, it didn't pass. All amendments need 60 votes for passage but stand alone bills need a simple majority.[7]

All Democrats voted for the amendment except Harkin of Iowa, Leahy of Vermont, Levin of Michigan, Nelson of Nebraska, Pryor of Arizona, Stabenow of Michigan and Webb of Virginia. Sanders, an Independent from Vermont also opposed the amendment.

All Republicans voted against the amendment except Burr of North Carolina, Chambliss of Georgia, Coburn of Oklahoma, Collins of Maine, Isakson of Georgia, and Snowe of Maine.

There were less than 6,000 medium and heavy duty natural gas engines that were estimated to have been sold during 2011. This data was reported by the two largest manufacturers: Cummins Westport Inc. (CWI) and Westport Innovative Inc. A relatively small number of those vehicles were shipped to Canada and China.[8]

Light to medium trucks, vans and buses can be converted to natural gas fuels and are proceeding at a faster pace in comparison to the heavy duty trucks. For example, BAF which is a subsidiary of Clean Energy, states that they have completed "more than 15,000 conversions for taxis, vans, pickup trucks and shuttle buses. With a full line of Ford vehicles, BAF is the first Qualified Vehicle Modifier (QVM) designated by Ford Motor Company for gaseous fuel."[9] Other companies including IMPCO have dedicated resources for manufacture and conversion to natural gas fuel engines.

There are many companies and government entities that are converting to natural gas. One such is Waste Management, which purchased its 1,000th natural gas vehicle and has plans to convert its entire refuse fleet to the new fuel in the future.[10]

California has a program where all new replacement refuse-trucks must be operating solely on natural gas.

HUGE MARKET

The potential market is huge. There are 8.7 million registered heavy duty trucks in the United States and a survey indicated that 50% of these vehicles travel 200 miles or less on a daily basis which is well within the driving range of natural gas vehicles.[11]

"While there are hundreds of trucking companies within the United States, there are only four domestic commercial truck producers and about half a dozen major manufacturers making up the North American market. Additionally, powertrain components are sourced from only about the same number of manufacturers. This means there is a lot of consolidation within the industry's supply chain. With no known exception, every major manufacturer of heavy-duty commercial vehicles is experimenting with or has production models that are utilizing fuels that are not primarily sourced from crude oil. Premier amongst these are natural gas and biofuels options."[12]

In comparison, natural gas engines for the trucking and transit industries are more advanced and meet standard driving requirements. If one million of these vehicles are on the road, then an estimated replacement value of eliminating more than 25 million cars occurs. This is based on trucks traveling 90,000 miles and getting six mpg or using 15,000 gallons per year; and on passenger cars traveling 15,000 miles and getting 25 mpg or using 600 gallons per year.

ISSUE OF ENGINE COST

The other half of this equation consists of how to expedite the purchase of hundreds of thousands of fleet and commercial vehicles on an economically profitable basis. Besides finding a location for fuel, there is the issue that the cost to purchase a medium to heavy duty natural gas truck engine is more than one using diesel by a range to $30,000 to $70,000 depending on the size of the vehicle. Only greater future orders will bring down costs through mass production.

Natural gas as a transportation fuel is about $1.50 to $2.00 less per equivalent gallon than diesel which allows the owner to earn back the increased cost for this type of engine. It is easily possible for a commercial vehicle to consume 15,000 gallons of diesel per year as described above.[13]

SAVINGS

The mileage depends chiefly on the weight of the truckload and the terrain. The owner or operator would save $22,500 to $30,000 each year as long as the spread or difference between natural gas and diesel remains $1.50 to $2.00. Most estimates of future costs of natural gas consists of mild to moderate increases from today's all time lows but a significant price difference will remain in most forecasts in comparison to diesel for many years.

END NOTES:

A companion web site has been established to go along with this report so that readers may view pictures, graphs and maps in color and link direct to referenced web sites.

The URL to this "Master" link is:

http://natgas-rpt.com

1 Energy Statistics, >Oil>consumption (most recent) by country (2009) NationMaster.com. Original source CIA World Factbooks December 18, 2003 to March 28, 2011. Retrieved from **http://www.nationmaster.com/graph/ene_oil_con-energy-oil-consumption**

2 U. S. & world population clocks. (2012, August 23) Current Population Clock, United States Census Bureau. Retrieved from **http://www.census.gov/3main/www/popclock.html**

3 Current natural gas vehicle statistics. (2011, December) NGV Global, Natural gas vehicle knowledge base. Retrieved from **http://www.iangv.org/current-ngv-stats/**

4 Current natural gas vehicle statistics. Ibid

5 Current natural gas vehicle statistics. Ibid

6 NGV Global statistics show irrepressible growth of NGVs. (2012, July 12) NGV Global, Natural gas vehicle knowledge base. Retrieved from **http://www.iangv.org/category/stats/**

7 U.S. Senate Roll Call Votes 112th Congress – 2nd Session, (2012, March 13) Vote Summary Question: On the Amendment (Menendez Amdt. No. 1782). Unites States Senate.gov.   Retrieved from **http://www.senate.gov/legislative/LIS/roll_call_lists/roll_call_vote _cfm.cfm?congress=112&session=2&vote=00041**

8 Westport Financial Report Fiscal Year 2011. Westport Innovations annual report for year ending December 31, 2011, Letter to Shareholders, p.2. Retrieved from **http://www.westport.com/investors/financial**

9 Clean Energy 2011 annual report, At a glance. Clean Energy 2011 annual report, At a glance, para. 8 BAF. Retrieved from **http://www. cleanenergyfuels.com/annualreport/2011/atglance.htm1**

10 Waste Management adds 1000th truck to natural gas fleet. (2011, July 12) Waste Management press release, About us, Press room, 2011. Retrieved from **http://www.wm.com/about/press-room/2011/20110712-wm-adds-1000th-natural-gas-truck.jsp**

11 Annual energy outlook 2010 with projections to 2035. Energy Information Administration (EIA), released: issues in Focus, AEO2010, p.34 last para. Retrieved from **http://www.eia.gov/ oiaf/aeo/otheranalysis/aeo_2010analysispapers/natgas_fuel.html**

12. Turpen A. (2012, April 16) Trucking industry moving to end reliance on crude, The time is right for commercial transportation to move away from petroleum and embrace new technologies. The shift has already begun, *Torque News*. Retrieved from **http://www.torquenews.com/1080/trucking-industry-moving-end-reliance-crude**

13. This is based on 400 miles a day times 250 days equals 100,000 miles a year at 6 miles per gallon. Current natural gas vehicles statistics 2011 as reported by Natural Gas Vehicles Global (NGV Global) **http://www.iangv.org/current-ngv-stats/**

CHAPTER XI
PUBLIC PARTNERSHIP

The ideal public/private partnership would be for the federal government to enact legislation to assist private enterprise in the conversion of natural gas as a fuel for transportation. As long as there are elements in Congress capable of blocking such legislation, this is not going to happen. Still there are other departments of the federal government that are promoting natural gas.

And there is still a place for a public/private partnership with the myriad of government agencies within each of the 50 states. California is a good example. The state is large in all aspects and has the most population in the country. California has 35 local Air Quality agencies, 53 Congressional districts, 58 counties, 40 state Senate Districts, 80 Assembly Districts, 1,131 school districts and 81 cities with a population of more than 100,000. Two cities, Los Angeles and San Diego have a population of more than one million.

The state has cast a wide net in encouraging alternative energy by creating 12 California areas that cooperate with the U.S. Department of Energy's Clean Cities Program.

The Clean Cities Program was established in 1993 under the Energy Policy Act. Today, it claims partnerships in nearly 100 cities.

It says, "In nearly 90 coalitions, government agencies and private companies voluntarily come together under the umbrella of Clean Cities. The partnership helps all parties identify mutual interests and meet the objectives of reducing the use of imported oil, developing regional economic opportunities and improving air quality."[1]

It goes on to say, "Clean Cities draws stakeholders from local, state and federal agencies; public health and transportation departments; commercial fleets; transit agencies; and other government offices; as well as auto manufacturers, car dealers, fuel and equipment suppliers, public utilities and nonprofit associations. More than 6,500 stakeholders have accepted the invitation to contribute to the partnership's mission."[2]

"Since its inception in 1993 Clean Cities and its stakeholders have displaced more than 2 billion gallons of petroleum."[3]

California cities and regions taking advantage of the financial and other assistance provided by the Department of Energy (DOE) are Antelope Valley (Lancaster), Central Coast (San Luis Obispo), Coachella Valley Region, East Bay (Oakland), Long Beach, Los Angeles, Sacramento, San Diego Region, San Francisco, San Joaquin Valley (Bakersfield), Silicon Valley (San Jose), and Western Riverside County.[4]

Long Beach is a good example of this partnership with the DOE. Projects include replacement of 180 diesel drayage trucks at the ports of Los Angeles and Long Beach that will displace 1.8 million gallons of petroleum annually.

Total DOE award was $9,408,000. An award of $5,591,600 by the DOE went toward funding "a long planned regional liquefied natural gas (LNG) corridor across the southwestern U.S., making the final connection between the existing public access LNG fuel infrastructure in Southern California and the LNG fuel stations being developed in Utah."[5]

CALIFORNIA GOVERNMENT

The California Senate and Assembly have participated in providing creative solutions to the energy problem by establishing a number of committees which promote natural gas as an alternative fuel for transportation. Most state's counties, major cities and larger school districts either have partnerships or have the opportunity to partner with public and non-public entities to promote the conversion of natural gas as an alternative fuel for transportation.

In Los Angeles both major transportation entities composed of the city and school district are fueled by natural gas. The Los Angeles public transportation system operates 2,228 buses which it claims as "the first major transit agency in the world to operate only alternative clean fueled buses."[6] "[For any school organization] ...the Los Angeles Unified School District operates the largest alternative fuel bus fleet in California of 475 compressed natural gas, 100 ultra low emission and 126 propane buses comprising over 50 percent of the bus fleet."[7]

PUBLIC PRIVATE PARTNERSHIPS

On a national basis, the federal government has been successful in past generations in creating public/private partnerships that have changed the way Americans live and have promoted economic prosperity.

In his book *Drill Here Drill Now Pay Less*, Newt Gingrich, former speaker of the House, puts forth a "Bold Plan for America's Energy Independence." He cites major accomplishments of the country during World War II as examples of how "We can do it all, we can do it now, we can do it for America."[8]

- America built a vast industrial infrastructure from scratch.
- It built the Pentagon in just 16 months

- "Cities to house hundreds of thousands were built from scratch" to house those working on the Manhattan project that developed the atomic bomb.

Before 1956 most highways in America were two and three lane. That year the Federal-Aid Highway Act was enacted and within five years most of America was connected by an Interstate Highway System.[9]

President Kennedy announced the goal of sending astronauts to the Moon and in six years after Kennedy's death, Neil Armstrong walked on the moon.

Public funding of roads led to private enterprises' tremendous growth in vehicles of all types and oil became further entrenched as the primary source of transportation fuel for America. Private enterprise became more efficient and costs were reduced in transporting the country's merchandise and in business travel in general.

TAX ADVANTAGES

There were also many tax advantages provided the oil industry in depreciating their exploration costs and equipment expenditures. In addition, tax deductions were allocated for people who used their vehicles for business purposes. Overall, this was an exemplary achievement in utilizing public and private cooperation for America's advantage. Another example consists of how the internet was developed by the federal government, opened up to the private sector and then subsidized by favorable taxation policies.[10]

The country wanted to assist the expansion of the fledgling internet businesses and exempted them from paying local and state taxes. This powerful economic advantage helped propel internet commerce into the mainstream of our lives. The industry is so strong now that it doesn't need the added benefit of not paying taxes. Therefore as of September 2012, internet purchases originating in California will be subject to sales and local taxes. There will be a push by all states to do the same.

Construction and maintenance of the national highway system and creation of the Internet are huge success stories involving public/private partnerships. Why can't the same pattern be applied to jump start the use of natural gas for transportation purposes?

END NOTES:
A companion web site has been established to go along with this report so that readers may view pictures, graphs and maps in color and link direct to referenced web sites

The URL to this "Master" link is:

http://natgas-rpt.com

1 What is clean cities? (2009, November) U.S. Department of Energy, Energy Efficiency & Renewable Energy, Clean Cities. p.1 para 2. Retrieved from **http://www1.eere.energy.gov/library/pdfs/46689.pdf**

2 How does clean cities work? (2009, November) U.S. Department of Energy, Energy Efficiency & Renewable Energy, Clean Cities. p.1 para 1. Retrieved from **http://www1.eere.energy.gov/library/pdfs/46689.pdf**

3 Mission statement, Clean Cities, Long Beach. City of Long Beach, CA, public works. Retrieved from **http://www.longbeach.gov/pw/longbeachcleancities/**

4 Clean Cities coordinators (2009, November) U.S. Department of Energy, Energy Efficiency & Renewable Energy, Clean Cities. p.2. Retrieved from **http://www1.eere.energy.gov/library/pdfs/46689.pdf**

5 LNG drayage truck project. Clean Cities, Long Beach. City of Long Beach, CA, public works. para 2-3. Retrieved from **http://www.longbeach.gov/pw/longbeachcleancities/**

6 Metro retired last diesel bus, becomes world's first agency to operate only clean fuel buses, (2011, January 12) Metro, News & media, news releases, para 1 bold. Retrieved from **http://www.metro.net/news/simple_pr/metro-retires-last-diesel-bus/**

7 Parents & students. Los Angeles Unified School District, Transportation Services Division, para. 4. Retrieved from **http://transportation.lausd.net/Parents_and_Students**

8 Gingrich, N. and Haley, V., (2008*) Drill Here Drill Now Pay Less,* Regnery Publishing Inc. Washington D. C.

9 Dwight D. Eisenhower national system of interstate and defense highways. (2012, March 14) U. S. Department of Transportation Federal Highway Administration, Design, Interstate System. Retrieved from **http://www.fhwa.dot.gov/programadmin/interstate.cfm**.

10 Internet history. Connected: An Internet Encyclopedia. Retrieved from **http://www.freesoft.org/CIE/Topics/57.htm**.

CHAPTER XII
NATURAL GAS ACT

Natural gas bills designed to promote natural gas as an alternative fuel for transportation were introduced in both houses of the 111th and 112th Congress. In none of the sessions of Congress did any bill make it out of committee. The bills did not have the backing of the oil companies.

The four bills introduced in 2009 and 2011 in the House and Senate have been referred to variously as the Natural Gas Act, Natural Gas Act of 2009 and 2011 and so on. In the official version of the Government Printing Office, it is referred to as "To amend the Internal Revenue Code of 1986 to encourage alternative energy investments and job creation." The short title is the "New Alternative Transportation to Give Americans Solutions Act of 2009 (2011)."

The Natural Gas Act of 2009 was introduced in the House of Representatives as HR 1835 on April 1. Sponsor of the bill was Dan Boren (D-OK). There were 146 co-sponsors. The last action on the bill listed by the Library of Congress was April 6, 2009 when it was referred to the Subcommittee on Energy and Development from the Committee on Energy and Commerce.[1]

The 2009 bill was introduced in the Senate as S 1408 Sponsor of the bill was Robert Menendez (D-NJ). There were seven co-sponsors. The bill was referred to the Senate Finance Committee. Chairman of the Senate Finance Committee was Max Baucus, (D-MT) The committee chair determines whether a bill will move past the committee stage.[2]

ACTIONS IN 2011

On April 6, 2011, the Natural Gas Act of 2011 was introduced in the House of Representatives as HR 1380 Sponsor of the bill was a Republican, John Sullivan of Oklahoma. The bill had 179 cosponsors. The bill has been referred to three committees and a subcommittee. The three committees are on Ways and Means, on Science, Space and Technology, and Energy and Commerce "for a period to be subsequently determined by the Speaker, in each case for consideration of such provisions as fall within the jurisdiction of the committee concerned." It was finally referred to the Subcommittee on Energy and Power a subcommittee of the Committee on Energy and Commerce.[3]

The 2011 bill was introduced in the Senate on November 15 as S1863. Sponsor of the bill was Robert Menendez (D-NJ) and five cosponsors. The bill was referred to the Committee on Finance.[4]

In March of 2012 there was an attempt in the Senate to attach the provisions of the Natural Gas Act as an amendment to the transportation bill. It failed by a 51 yes to a 47 no vote.[5] A list of those voting for and against the bill is in Chapter X.

In all four bills introduced in 2009 and 2011, there were incentives to purchase natural gas fueled heavy duty trucks as well as personal passenger vehicles.[6-9] A copy of each of the two bills of 2011 is in Appendix E.[10-11] A copy of the two bills introduced in 2009 may be found at the Government Printing Office.[12, 13] A comparison of the two bills can be viewed on the web site constructed by National Gas Vehicles for America (NGV America) which is a national trade association.[14]

Provisions of the House Bill introduced in 2011 include:

1. Tax credits to private industry comprised of 89% of the differential between the purchase of a natural gas engine compared to a diesel model. There is a maximum benefit of $64,000 depending on the weight/size of the truck. This would provide a level playing field in relation to the original expense in purchasing a diesel vs. natural gas truck. The owner will immediately save about $22,500 per year in fuel costs depending on how many miles the truck is driven. If the owner has to pay for the increased cost on his own then the return on investment (ROI) will take about two to three years again dependent on mileage factors and the cost of the natural gas vs. diesel fuel.

2. Credits on personal passenger cars of 80% of the differential cost of purchasing a natural gas engine with a maximum benefit of $7,500. For installation of home fueling system, a $2,000 tax credit goes towards installation such a system.

3. A 50-cent per gallon fuel tax credit is in place in 2011.

4. Infrastructure tax credits of 50% of the cost of constructing a natural gas fueling station with a maximum benefit of $100,000. Current law has a ceiling of $30,000

5. A tax credit to the manufacturer for the production of natural gas vehicles.

There is a five year life span of the bill and estimated cost for the bill was $5 billion.

COSPONSORS

The legislation claimed to have 181 cosponsors at time of submission but the support dwindled as conservative backlash to the bill gathered steam. Institutes like the Heartland pummeled Congress and the public with opposing arguments against the bill.[15]

END NOTES:
A companion web site has been established to go along with this report so that readers may view pictures, graphs and maps in color and link direct to referenced web sites.

The URL to this "Master" link is:

http://natgas-rpt.com

1 H.R. 1835, All Congressional actions. Bill summary & status 111th Congress (2009-2010). Library of Congress, Thomas. Retrieved from **http://thomas.loc.gov/cgi-bin/bdquery/z?d111:HR01835:@@@X|/ home/LegislativeData.php?n=BSS;c=111| All congressional actions on HR 1835 of 2009**

2 S. 1408, All Congressional actions. Bill summary & status, 111th Congress (2009-2010). Library of Congress, Thomas. Retrieved from **http://thomas.loc.gov/cgi-bin/bdquery/z?d111:SN01408 :@@@X All congressional action on S 1408 of 2009**

3 H.R. 1380, All Congressional actions. Bill summary & status, 111th Congress (2009-2010). Library of Congress, Thomas. Retrieved from **http://thomas.loc.gov/cgi-bin/bdquery/z?d112:HR0 1380:@@@X All congressional actions on HR 1380 of 2011**

4 S. 1863, All Congressional actions, Bill summary & status, 112th Congress (2011-2012). Library of Congress, Thomas. Retrieved from_ **http://thomas.loc.gov/cgi-bin/bdquery/z?d112:SN01863 :@@@X All congressional actions on S 1863 of 2011**

5. U.S. Roll call votes 112th Congress – 2nd session, Vote Summary: On the Amendment (Menendez Amdt. No. 1782). United States Senate.gov. Retrieved from **http://www.senate.gov/legislative/LIS/roll_call_lists/roll_call_vote_cf m.cfm?congress=112&session=2&vote=00041**

There is an official description of each of the following bills at:

6 HR 1835 (111th): New alternative transportation to give Americans solutions act of 2009 Retrieved from **http://www.govtrack.us/congress/bills/111/hr1835**

7 S 1408 of 2009 (111th): New alternative transportation to give Americans solutions. Retrieved from **http://www.govtrack.us/congress/bills/111/s1408**

8 HR 1380 New alternative transportation to give Americans solutions act of 2011. Retrieved from **http://www.govtrack.us/congress/bills/112/hr1380**

9 S 1863: New alternative transportation to give Americans solutions act of 2011. Retrieved from **http://www.govtrack.us/congress/bills/112/s1863**

Copies of the following first two bills are in Appendix E. Copies of all four bills may be accessed from their respective links to the Government Printing Office.

10 HR 1380, To amend the Internal Revenue Code of 1986 to encourage alternative energy investments and job creation. (2011, April 6). 112th Congress first session. Retrieved from **http://www.gpo.gov/fdsys/pkg/BILLS-112hr1380ih/pdf/BILLS-112hr1380ih.pdf**

11 S 1863: New alternative transportation to give Americans solutions act of 2011. Govtrack.us, 112th Congress, 2011-2012. Retrieved from **http://www.govtrack.us/congress/bills/112/s1863**

12 H.R. 1835 To amend the Internal Revenue Code of 1986 to encourage alternative energy investments and job creation. (2009, April 1). 111th Congress. Retrieved from **http://www.gpo.gov/fdsys/pkg/BILLS-111hr1835ih/pdf/BILLS-111hr1835ih.pdf**

13 S. 1408 To amend the Internal Revenue Code of 1986 to encourage alternative energy investments and job creation. (2009, July 8). 111th Congress. Retrieved from **http://www.gpo.gov/fdsys/pkg/BILLS-111s1408is/pdf/BILLS-111s1408is.pdf**

14 Nat Gas Acts – S 1408-vs. HR 1835 – Side by Side. (2009, July 21). NGV America, Natural gas vehicles for America. Retrieved from **http://www.ngvc.org/pdfs/S1408vsHR1835_NATGAS 111th_SidebySide_072109.pdf**

15. Letter of opposition: Americans For Prosperity (AFP) rejects natural gas act energy subsidies, H.R. 1380. (2011, May 10).   Retrieved from **http://americansforprosperity.org/legislativealerts/051011-letter-opposition-afp-rejects-nat-gas-act-energy-subsidies-hr-1380/**e

CHAPTER XIII
WHAT IF...

What if there were an abundance of natural gas refueling stations and none less than 100 to 200 miles apart on all Interstate highways.

What if there were no significant price differential between a natural gas vehicle and a comparable petroleum fueled vehicle?

What if city, county, state and federal governmental agencies were required to purchase natural gas fueled vehicles.

If this were to happen...

Within a few years, there would be hundreds of thousands of natural gas fueled vehicles on the road.

There would be increased competition resulting in substantial downward pressure on the price of gasoline and diesel.

The trade deficit attributed to the importation of foreign oil would be substantially decreased if not altogether eliminated.

The export of petroleum products would substantially increase.

The annual injection of hundreds of billions of dollars into the economy that would otherwise be sent overseas would stimulate economic activity and cause the creation of hundreds of thousands of jobs.

The air would be cleaner.

There would be less money funding terrorism.

IT CAN HAPPEN

It can easily happen. All that is lacking is political will. To have natural gas refueling stations 200 miles apart (or closer) on all Interstate highways, there would need to be an additional thousand or more new stations built. If the federal government stepped in this could be accomplished in three to five years and create hundreds of thousands of new jobs.

There are many ways to equalize the playing field between the price of a natural gas powered vehicle and a petroleum fueled vehicle. All it takes is political will and it need not add one cent to the deficit. There is plenty of room between the price of natural gas and gasoline and diesel to make up the difference in new taxes, tax exemptions, rebates, loan repayment and other programs to provide incentives to purchase a natural gas vehicle.

GOVERNMENT VEHICLES

In many states there are thousands of vehicles a year purchased by governmental agencies. This could easily add thousands of natural gas vehicles a year to the number on the road if governmental agencies across the nation were required to purchase them rather than conventional fueled vehicles.

Currently the two most popular manufacturers of heavy duty natural gas truck engines in North America (Cummins Westport Inc. and Westport Innovative) produced less than 6,000 units for all of 2011.[1] This figure becomes the baseline for considering future progress.

What if there were an effort to purchase and/or convert to natural gas a combined total of 100,000 medium to heavy duty trucks on a yearly basis. This achievement would also involve construction of an adequate infrastructure to support these vehicles. Only through a partnership effort would this goal be achievable within the five year time period.

The target of 100,000 natural gas trucks amounts to 1.3 percent of oil imports from all countries outside of North America. So after 10 years of purchasing and converting at the same level (100,000 trucks annually), the United States would achieve about a 13 percent reduction. These calculations are based on replacing natural gas fuel for medium and heavy duty trucks which travel 90,000 or more miles annually and consume 15,000 gallons of diesel.

The 100,000 trucks operating on natural gas will replace 4.4 percent of the oil imported from the Middle East. If the same or greater rate of conversion was continued for 10 years, this would be a 44 percent reduction.

The target figure of purchasing 100,000 natural gas trucks represents approximately 30 percent of market share of all medium and heavy duty vehicles sold in normal year of financial activity.

For example in a normal year like 2007, there were sales totaling more than 371,000 trucks, but in 2009 at the depth of the Great Recession, there were sales totaling only 199,700 vehicles. Substantial recovery was evident last year in 2011 when more than 306,000 units were purchased according to Wards Automotive Research.[2]

The above target of 100,000 units does not include using natural gas for operating any light duty vehicles such as pickup trucks, taxis and mini-vans. In addition, there are plans in a few years for producing locomotives by GE and earth-moving equipment by Caterpillar that would operate on natural gas as well.

These high volume diesel-eating machines would result in a greater reduction of oil within America. Therefore the above reductions over ten years for 13% (all countries outside of North America) and 44% (Middle East region) are conservative depending on how rapidly these new products are brought to market.

Continued progress in purchasing high mileage passenger cars including the impact of future electric vehicles, would yield about an equivalent reduction in gasoline fuel over the same ten year period. This amounts to another 13% reduction from all countries and 44% elimination from Middle East region if all fuel cutbacks were allocated to that region.

The basis of this estimate relies on the significant improvement of new CAFE fuel consumption standards by the transportation industry and government. For example, highlights of the combined "estimated average required for fleet-wide fuel economy" for both cars and trucks are summarized as follows.[3]

Base year of 2016 = 34.1 mpg
2020 = 38.8 mpg, 14% less fuel compared to base year;
2025 = 49.6 mpg, 45% less fuel compared to base year.

SAVINGS

Adding up the reductions from natural gas trucks and high mileage passenger cars means that substantial savings are achieved over a ten year period. Oil imports from all countries outside of North America are reduced by 26 percent but by targeting only the Middle East region, there is an 88 percent reduction in oil imports. Thus energy independence from the most volatile region of the world and from countries who want to do America harm is achieved.

The economic benefits within the United States will mean tens if not hundreds of billions of dollars of increased financial activity. Estimates of 600,000 plus new jobs have been calculated by NGV America (Natural Gas Vehicles for America and Pickens Plan).[4] Transportation costs in general by heavy-duty trucks would decline since natural gas is much less expensive than diesel. This results in increased profits for the transportation sector given the lower cost of natural gas.

The manufacturing of organic or carbon based products such as plastics and fertilizer would also expand due to the inexpensive and large supply of natural gas which is used as a feedstock for these products. The chemical industry states that hundreds of thousands of jobs would be created, and again the American economy would expand by many billions of dollars in economic activity.[5]

Major oil companies are making acquisitions in natural gas. ExxonMobile acquired XTO Corporation in 2010 which was the largest natural gas company in U.S. when it was purchased.[6] In addition, Shell Oil bought East Resources, another coveted company that had extensive natural gas resources.[7]

END NOTES:

A companion web site has been established to go along with this report so that readers may view pictures, graphs and maps in color and link direct to referenced web sites.

The URL to this "Master" link is:

http://natgas-rpt.com

1 Letter to shareholders in Annual Report for the year ending December 31, 2011. Westport Innovative, Financial
information, page 2. Retrieved from
http://www.westport.com/investors/financial

2 See APPENDIX B Comparative Factors of replacement of imported oil by using natural gas and high mileage passenger vehicles.

114

3 NHTSA and EPA Propose to Extend the National Program to Improve Fuel Economy and Greenhouse Gases for Passenger Cars and Light Trucks, (2011, November), Office of Transportation and Air Quality Environmental Protection Agency, Regulatory announcement. http://www.epa.gov/oms/climate/documents/420f11038.pdf

4 Natural gas vehicles mean more American jobs and a stronger economy. Natural Gas Vehicles for America (NGV American) and Chesapeake Energy. Retrieved from http://www.cngnow.com/Tagged%20PDFs/CHK_Economy.pdf

5 Shale gas and new petrochemicals investment: Benefits for the economy, jobs and US manufacturing. (2011, March). Economic & Statistics American Chemistry Council, executive summary and p.26 & 27, Tables 2 & 3. Retrieved from http://www.americanchemistry.com/ACC-Shale-Report

6 Yousef, H. (2009, December 14) Exxon to buy XTO in 441 billion deal, Biggest U.S.oil company seeks to expand its role in the natural gas industry. *CNN Mone.comy.* Retrieved from http://money.cnn.com/2009/12/14/news/companies/Exxon_Mobil_XTO/index.htm.

7 Pals, F. (2010, May 28) Shell buys U.S. gas assets from East Resources for $4.7 billion. *Bloombnerg* Retreived from http://www.bloomberg.com/news/2010-05-28/shell-agrees-to-buys-subsidiaries-of-east-resources-for-4-7-billion.html

8. Roll call vote on the amendment in the U.S. Senate. (senate 2) http://www.senate.gov/legislative/LIS/roll_call_lists/roll_call_vote_cfm.cfm?congress=112&session=2&vote=00041

9. Barr, A.(2010, October 28), Politico "John Boehner: 'We will not compromise'", (politico 2) http://www.politico.com/news/stories/1010/44311.html#ixzz15Tkg0d9b

CHAPTER XIV
LIGHT DUTY VEHICLES

Buying or converting cars, small pickups and vans to natural gas can save money for individuals and businesses. Most of this report has been concerned with medium and heavy duty vehicles, mainly trucks and buses, the conversion of which to natural gas, would have the greatest impact on the environment, stopping the importation of foreign oil and defunding terrorism. This is really only half the story.

The case needs to be made for the average driver converting to natural gas.

The cost of natural gas in California during the summer of 2012 was running about $1.50 per gge (gasoline gallon equivalent). You can't fill up your natural gas vehicle for that price. That's the general price Southern California Gas Company is charging its customers to heat their homes, heat their water and to cook with.

To fill up your vehicle's tank with natural gas, the natural gas has to go through a conversion. The two kinds of conversions of natural gas most in use are compressed natural gas (CNG) and liquefied natural gas (LNG). Compressed natural gas has to be compressed and liquid natural gas has to be frozen. Right now, CNG is the most popular fuel for light duty vehicles.

For CNG to be used it has to be compressed to 3000 pounds per square inch (psi) or 3600 psi. You can buy a compressor and fill up at home or you can go to a compressed natural gas fueling station. To fill up at one of the many fueling stations in Southern California the summer of 2012, would have cost from $2 to $3 a gge. Honda is said to be the only auto company making a natural gas passenger car and it gets about 25 gge.. The Honda natural gas tank holds eight gallons of gge of natural gas. So to fill up a Honda will cost between $16 and $24 versus more than $32 for gasoline. If you are in Oklahoma City, you can fill up for $10, or $1.29 a gge.

MOST NGV'S ARE LIGHT DUTY

There is possibly around 124,000 natural gas vehicles on the road in the U.S. More than half are light duty vehicles, possibly as many as 65,000. As is illustrated in Appendix "B," the market for light duty natural gas vehicles is 20 times that of medium and heavy duty vehicles.

Social movements usually start with the public and ultimately result in Congressional action. While it is the conversion to natural gas of medium and heavy duty trucks and buses that will make the major difference in stopping the importation of foreign oil, it is the owners of light duty natural gas vehicles—passenger cars, SUVs, pickups and vans—who likely may ultimately be the determining factor in the U.S. Congress passing a natural gas act, which will result in a substantial increase in the number of fueling stations.

At some point in the near future, the public is going to realize the value of converting to natural gas. It will be the general public that will create the demand for more natural gas fueling stations, which in turn will promote the sale of more heavy duty natural gas trucks.

What may propel the use of natural gas forward is the development of a more efficient home fueling system. Right now home fueling systems cost from $7,000 to $10,000 to install. They take from 8 to 16 hours to fill up a Honda, depending on the model of fueling system. It takes only a few minutes to fill up a compressed natural gas tank of a vehicle at established commercial fueling stations.

Now imagine having a home fueling system in your garage connected to your gas line You plug it into your vehicle and have a full tank of CNG within an hour. And the fueling system costs less than $500. According to the web site, *Shale Stuff*, General Electric along with Chart Industries and scientists at the University of Missouri are working to develop just such a system. Cost to install was not given. Cost to install current systems is usually more than $1500. The groups were awarded a program through Advanced Research Projects Agency for Energy (ARPA-E).[1]

The *Shale Stuff* web site reported "The research team from GE, Chart Industries, and the University of Missouri will design a system that chills, densifies, and transfers compressed natural gas... It will be a much simpler design with fewer moving parts, and that will operate quietly and be virtually maintenance-free."[2]

We have talked to a number of owners of natural gas vehicles in the Los Angeles area and they have generally expressed satisfaction with their vehicles performance and fuel availability. Truck drivers have complained of a lack of power, especially going up hill. Honda Civic owners only complaint was handiness of fueling stations.

Drivers of alternative fuel vehicles are allowed in all lanes of California freeways. Honda owners said they averaged savings of a half hour to an hour on their daily commutes by using the diamond lane.

END NOTES:
A companion web site has been established to go along with this report so that readers may view pictures, graphs and maps in color and link direct to referenced web sites.

The URL to this "Master" link is:

http://natgas-rpt.com

1. Ferguson, T. (2012, July 27) GE developing at-home refueling system for NG vehicles. *Shale Stuff.* Retrieved (September) from **http://shalestuff.com/environment-2/ge-developing-at-home-refueling-station-ng-vehicles/article02257**

2. Ferguson, T. Ibid

APPENDIX A
EPILOGUE

Compiling a report of this magnitude involves a tremendous amount of effort. It seems the more information that is uncovered, the more there remains to be uncovered. Research is like that. One thing leads to another thing, which leads to another thing and so on. This is what we have found in compiling this report. We are afraid we have just scratched the surface. This preliminary edition is published so that those involved can provide feedback and information that can be included in future printings.

We have set up a companion web site, which includes supplements and updates as well as End Notes to each chapter so that readers can connect direct to the active URLs that are referenced.

The U.S. prides itself on having a free-market economy driven by competition. That's the basic definition of capitalism. However, much of the U.S. economy is subverted by businessmen seeking to gain an advantage over their competition or by monopolizing a certain segment of the economy and they are aided and abetted by lawmakers who are beholden to special interests.

We intend to continue our investigations, which will be reported in the supplements of the companion web site and which at some future date will be included in a second edition.

We are appreciative of those who have contributed to this report and will continue to be appreciative of those in the future, who provide us with additional information. As usual we will affirm our commitment to keep confidential sources of information who do not want to be identified.

APPENDIX B
COMPARATIVE FACTORS

The following series of tables display sales of the eight classes of vehicles on the road in the United States and the effect conversion of a certain number of heavy duty vehicles to natural gas would have on oil imports. The eight different classes of vehicles are listed and explained in Chapter VIII. Classes #1, #2 and #3 are light duty vehicles such as passenger cars and small pickups. Classes #4, #5 and #6 are medium duty trucks and buses. Classes #7 and #8 are heavy duty trucks such as tractors and dump trucks.

Weight by Class	2007	2008	2009	2010	2011
Light Duty	8,526,888	6,425,634	5,000,792	5,9129,085	6,845,021
Medium Duty	220,128	164,951	104,888	110,550	134,831
Heavy Duty	150,965	133,473	94,798	107,152	171,358
Totals: Medium & Heavy Duty	371,093	298,424	199,686	217,702	306,189

Source: , Ward's Automotive Group, a division of Penton Media Inc. (WardsAuto.com, copyright 2012) Used with permission.

Medium and Heavy Duty Truck Sales Summary:

5 Year Total sales	**1,393,094**
5 Year Average sales years	**278,619**
3 Year Total sales of trucks 2007, 2010 & 2011	**894,328**
3 Year Average sales of trucks 2007, 2010 & 2011*	**298,328**

*(Excludes sales of worst years of recession)

Truck sales for medium and heavy duty trucks varied from a low of 200,000 in 2009 at the depth of the Great Recession to a high of 371,000 prior to the recession. Natural gas trucks of all weights and classes need to be more than 100,000 per year in order to make a significant impact in reducing U.S. consumption of foreign imported oil (oil originating from countries outside of North America). This means that a market penetration of 30% is necessary per year during normal sales activity of 300,000 new trucks. This doesn't count reduction on petroleum usage from SUV's or passenger vehicles.

2011 IMPORTED OIL DATA

U.S. imports of petroleum from country of origin

Data source: U. S. Energy Information Agency (August 2012). Retrieved from **http://www.eia.gov/dnav/pet_ move_impcus_a2_nus_ep00_im0_mbbl_a.htm**

4,126,266,000	Total barrels of oil imported from ALL foreign countries
174,143,172,000	Total gallons (barrels multiplied by 42 gallons.per barrel)
987,736, 000	Total barrels of oil imported from Canada
436,753,000	Total barrels of oil imported from Mexico
1,424,489,000	Total barrels of oil imported from Canada and Mexico
2,721,177,000	Total barrels of oil imported from outside of North America. (subtracted oil from Canada & Mexico)
114,314,634,000	Total gallons of oil imported from outside of North America (barrels multiplied by 42 gallons per barrel)

2011 MIDDLE EAST ONLY OIL IMPORTS

Barrels imported	Country
130,558,000	Algeria
167,905,000	Iraq
69,880,000	Kuwait
5,462,000	Libya
2,158,000	Qatar
436,051,000	Saudi Arabia
3,645,000	United Arab Emirates
815,669,000	Total imported oil from Middle East in barrels
34,258,098,000	Total gallons imported from Middle East for 2011 (barrels multiplied by 42 gallons per barrel

ESTIMATED IMPACT FOR USING NATURAL GAS TO OPERATE MEDIUM AND HEAVY DUTY TRUCKS IN U.S.

100,000	Natural gas trucks purchased &/or converted annually
15,000	Number of gallons of diesel each truck consumes given 90,000 miles @ 6 mpg
1,500,000,000	Number of gallons of diesel replaced annually by using natural gas
114,314,634,000	Total number of gallons of oil imported annually from outside of North America in 2011
1.31%	Percentage of replaced fuel compared to oil imported from outside of North America (1,500,000,000 / 114,314,634,000)
13.1%	Percentage of replaced diesel fuel for 10 years if maintain purchase / conversion of 100,000 trucks
34,258,098,000	Total gallons imported from Middle East for 2011
4.4%	Percentage of replaced diesel fuel if imported oil is targeted solely from Middle East
44%	Percentage of replaced diesel fuel if imported oil is targeted from Middle East for 10 years

Estimate that an equivalent amount of gasoline is replaced by the purchase of high mileage passenger vehicles given implementation of CAFÉ standards. CAFE data Retrieved from **http://www.nhtsa.gov/fuel-economy/**

1.31%	Percentage of replaced gasoline per year when compared to all imports – outside of North America
13.1%	Replaced gasoline for 10 years when compared to all imports – outside of North America
4.4%	Percentage of replaced diesel fuel compared to amount of oil imported from Middle East
44%	Percentage of replaced diesel fuel if the purchase of high mileage passenger cars is maintained for 10 years

TOTAL SAVINGS OF IMPORTED OIL FROM NATURAL GAS TRUCKS AND HIGH MILEAGE CARS

Replacement of Middle East oil for 10 years amounts to 44 percent for Trucks plus 44 percent for Cars. This in turn amounts to an 88 percent reduction from the 2011 consumption level. Replacement of oil from all foreign countries outside of North America for 10 years amounts to 13 percent for Trucks and 13 percent for Cars. This in turn amounts to a 26 percent reduction from 2011 consumption level from all foreign countries.

APPENDIX C
ACKNOWLEDGMENTS

We would like to acknowledge all those who have been helpful in compiling information for this report and who have contributed and are contributing to the editorial process.

Will Hoppins for design of book cover

Raymond Learsy for granting permission to quote from his book, *Oil & Finance The Epic Corruption Continues*

Clinton Harriman and Guy D. Cusumano for helping us understand what the natural gas industry is all about.

Energy Recovery Associates – for information on distributed LNG that can be utilized in passenger cars.

Lisa E. Williamson (Smith), WardsAuto.com, a publication of Ward's Automotive Group, a division of Penton Media Inc., for data on number of trucks purchased per year by classification.

Fred Duval, Consultant, Clean Energy, for valuable resource material and advice.

Rich Kolodziej, President of Natural Gas for Vehicles in America (NGV America) for valuable resource material and advice.

Paul Kerkhoven, Natural Gas for Vehicles in America (NVG America) Provided valuable resource material and advice.

Janet Cohen, USEPA Office of Transportation and Air Quality, Reviewed material on EPA standards pertaining to conversion to natural gas engines.

Lila Silvern for her critical insights.

APPENDIX D
DEFINITIONS

AFV Alternative fuel vehicle

Biogas A gas usually composed of methane gas (50% to 80%) and carbon dioxide (50% to 20%) that is produced from natural processes through bacterial degeneration of organic material; method is described as anaerobic digestion i.e. decomposition without oxygen. The gas can be used to generate electricity, heat and also utilized as a transportation fuel. Landfills and livestock operations are two major sources of biogas that originate from human activities.

Butane One of a group of gases characterized under. the category of natural gas; is a by-product of natural gas processing and petroleum refining. It can be liquefied and compressed.

CARB California Air Resources Board

CAFE Corporate Average Fuel Economy, mileage standards per gallon of fuel that EPA establishes for auto manufacturers who sell vehicles in the United States

CEC California Energy Commission

CNG Compressed natural gas. Natural gas that is compressed within an enclosure such as a tank that has reinforced walls to withstand pressures ranging from 2,900 to 3,600 pounds per square inch.

DLNG Distributed Liquefied Natural Gas. DLNG is methane liquefied into LNG at the point of use.

DOE U.S. Department of Energy

EER Energy economy ratio

EERE Office of Energy Efficiency and Renewable Energy

EIA Energy Information Administration

EPA Environmental Protection Agency

Ethane A colorless, odorless gas that occurs as a constituent of natural gas and is used as a fuel, a refrigerant and in the manufacture of organic chemicals

Ethylene A colorless flammable gas derived from natural gas and petroleum and used as a source of many organic compounds, in welding and cutting metals, to color citrus fruits, as an anesthetic, to manufacture plastics and fertilizers and as a fuel.

FHWA Federal Highway Administration

GGE, Gasoline gallon equivalent. The amount of energy that is produced by a gallon of gasoline. This value is used to equate gasoline to natural gas. Discussions in the text that compare amounts of gasoline and diesel to natural gas are based on gge or gallon of gasoline equivalent energy. One gge of gasoline equals 126.67 cubic feet of compressed natural gas and 1.52 gallons of liquefied natural gas.

GHG Greenhouse gas

HDV Heavy duty vehicle

LNG Liquefied natural gas. Natural gas that has been chilled to negative 260 degrees Fahrenheit. For transportation purposes, LNG enables vehicles to travel much greater distances due to the increased density and energy of the liquid.

LPG Liquefied petroleum gas gel. Propane that has been treated by mixing a number of benign chemicals with sand to create a thick gel that is then injected into gas wells to fracture shale formations. It enables companies to fracture without water which is a major environmental benefit.

NGV Natural gas vehicle

NREL National Renewable Energy Laboratory

PM Particulate matter

Propane One of a group of gases characterized under the category of natural gas; is a by-product of natural gas processing and petroleum refining; commonly used as fuel for engines, barbeques, portable stoves and heating; can be liquefied and compressed

VOC Volatile organic compounds

Wahhabism A form of Islam popularized in Saudi Arabia that has about a 200 hundred year history; favors a very conservative approach to Islam that shuns many modern societal practices and has created radical behavior in reaction to current social trends; violence has been associated with its implementation.

APPENDIX E
THE NATURAL GAS ACT

A natural gas bill has been introduced in the U.S. House of Representatives and in the U.S. Senate in 2009 and 2011.

The two bills introduced in 2011 were HR 1380 and S 1863. Following are copies of these two bills. Copy of Senate Bill 1863 starts on Page 155

APPENDIX E
NATURAL GAS ACT

Following are copies of HR 1380 and S1863 of 2011.

[Congressional Bills 112th Congress]
[From the U.S. Government Printing Office]
[H.R. 1380 Introduced in House (IH)]

112th CONGRESS
1st Session

H. R. 1380

To amend the Internal Revenue Code of 1986 to encourage alternative energy investments and job creation.

IN THE HOUSE OF REPRESENTATIVES

April 6, 2011

Mr. Sullivan (for himself, Mr. Boren, Mr. Larson of Connecticut, Mr. Brady of Texas, Mr. McCaul, Ms. Sutton, Mr. Gene Green of Texas, Mr. Shuster, Mr. Simpson, Mr. Bachus, Mr. Alexander, Mr. Grimm, Mr. Burton of Indiana, Mr. Thompson of Pennsylvania, Mr. Lujan, Mr. Critz, Mr. Bishop of Georgia, Mr. Cuellar, Mr. Doyle, Ms. Kaptur, Mr. Kissell, Mr. Lipinski, Mr. Matheson, Mr. Murphy of Connecticut, Mr. Ross of

Arkansas, Mr. Lucas, Mr. Welch, Mr. Cole, Mr. McIntyre, Mr. Bilbray, Mr. Culberson, Mrs. Blackburn, Mr. Donnelly of Indiana, Mr. Boustany, Mr. Fleming, Mr. Chandler, Mr. Hall, Mrs. Capito, Mr. Jones, Mr. Murphy of Pennsylvania, Mr. Rogers of Alabama, Mr. Perlmutter, Mr. Altmire, Mr. Gardner, Mr. Conaway, Mr. Ryan of Ohio, Mr. Sessions, Mr. Holt, Mr. Tonko, Mr. Sablan, Mr. Peters, Ms. DeGette, Mr. Capuano, Mr. Courtney, Mr. Clay, Mr. Thompson of California, Mr. Loebsack, Mr. Barton of Texas, Mr. Issa, Mr. Gallegly, Mr. Harper, Mr. Bishop of Utah, Mr. Terry, Mr. Costa, Mr. Barrow, Ms. Fudge, Mr. Cleaver, Mr. Serrano, Mr. Wu, Mr. Pascrell, Mr. Scalise, Mrs. Bono Mack, Mr. Boswell, Mrs. Lummis, Mr. Lankford, Mr. Rehberg, and Mr. Marchant) introduced the following bill; which was referred to the Committee on Ways and Means, and in addition to the Committees on Science, Space, and Technology and Energy and Commerce, for a period to be subsequently determined by the Speaker, in each case for consideration of such provisions as fall within the jurisdiction of the committee concerned

---

A BILL

To amend the Internal Revenue Code of 1986 to encourage alternative energy investments and job creation.

Be it enacted by the Senate and House of Representatives of the United States of America in Congress assembled,

SECTION 1. SHORT TITLE, ETC.

(a) Short Title.--This Act may be cited as the ``New Alternative Transportation to Give Americans Solutions Act of 2011".

(b) Amendment of 1986 Code.--Except as otherwise expressly

provided, whenever in this Act an amendment or repeal is expressed in terms of an amendment to, or repeal of, a section or other provision, the reference shall be considered to be made to a section or other provision of the Internal Revenue Code of 1986.

(c) Table of Contents.--The table of contents for this Act is as follows:

Sec. 1. Short title, etc.

TITLE I--PROMOTE THE PURCHASE AND USE OF NGVS WITH AN EMPHASIS ON HEAVY-DUTY VEHICLES AND FLEET VEHICLES

Sec. 101. Modification of alternative fuel credit.
Sec. 102. Extension and modification of new qualified alternative fuel motor vehicle credit.
Sec. 103. Allowance of vehicle and infrastructure credits against regular and minimum tax and transferability of credits.
Sec. 104. Modification of credit for purchase of vehicles fueled by compressed natural gas or liquefied natural gas.
Sec. 105. Modification of definition of new qualified alternative fuel motor vehicle.
Sec. 106. Providing for the treatment of property purchased by Indian tribal governments.

TITLE II--PROMOTE PRODUCTION OF NGVS BY ORIGINAL
EQUIPMENT MANUFACTURERS

Sec. 201. Credit for producing vehicles fueled by natural gas or
            liquefied natural gas.
Sec. 202. Amendment to section 136 of the Energy Security and
            Independence Act of 2007.

TITLE III--INCENTIVIZE THE INSTALLATION OF NATURAL GAS FUEL
PUMPS

Sec. 301. Extension and modification of alternative fuel vehicle
            refueling property credit.
Sec. 302. Increase in credit for certain alternative fuel vehicle
            refueling properties.

TITLE IV--NATURAL GAS VEHICLES

Sec. 401. Grants for natural gas vehicles research and development.
Sec. 402. Sense of the Congress regarding EPA certification of NGV
            retrofit kits.
Sec. 403. Sense of the Congress regarding EPA and NHTSA regulation of
            medium- and heavy-duty engines and
            vehicles.
Sec. 404. Amendment to section 508 of the Energy Policy Act of 1992.

TITLE I--PROMOTE THE PURCHASE AND USE OF NGVS WITH AN
EMPHASIS ON HEAVY-DUTY VEHICLES AND FLEET VEHICLES

SEC. 101. MODIFICATION OF ALTERNATIVE FUEL CREDIT.

(a) Alternative Fuel Credit.--Paragraph (5) of section 6426(d) (relating to alternative fuel credit) is amended by inserting ``, and December 31, 2016, in the case of any sale or use involving compressed or liquefied natural gas'' after ``hydrogen''.

(b) Alternative Fuel Mixture Credit.--Paragraph (3) of section 6426(e) is amended by inserting ``, and December 31, 2016, in the case of any sale or use involving compressed or liquefied natural gas'' after ``hydrogen''.

(c) Payments Relating to Alternative Fuel or Alternative Fuel Mixtures.--Paragraph (6) of section 6427(e) is amended--

   (1) in subparagraph (C)--

      (A) by striking ``subparagraph (D)'' and inserting ``subparagraphs (D) and (E)'', and

      (B) by striking ``and'' at the end thereof,

   (2) by striking the period at the end of subparagraph (D) and inserting ``, and'', and

   (3) by inserting at the end the following:

      ``(E) any alternative fuel or alternative fuel mixture (as so defined) involving compressed or liquefied natural gas sold or used after December 31, 2016.''.

(d) Payments Relating to Indian Tribes.--Paragraph (1) of section 6427(k)(A) is amended by inserting striking ``or'' at the end and inserting ``an Indian Tribal Government, or''.

(e) Effective Date.--The amendments made by this section shall apply to fuel sold or used after the date of the enactment of this Act.

SEC. 102. EXTENSION AND MODIFICATION OF NEW QUALIFIED ALTERNATIVE FUEL MOTOR VEHICLE CREDIT.

(a) In General.--Paragraph (4) of section 30B(k) (relating to termination) is amended by inserting ``(December 31, 2016, in the case of a vehicle powered by compressed or liquefied natural gas)'' before the period at the end.

(b) Effective Date.--The amendment made by subsection (a) shall apply to property placed in service after the date of the enactment of this Act.

SEC. 103. ALLOWANCE OF VEHICLE AND INFRASTRUCTURE CREDITS AGAINST REGULAR AND MINIMUM TAX AND TRANSFERABILITY OF CREDITS.

(a) Business Credits.--Subparagraph (B) of section 38(c)(4) is amended by striking ``and'' at the end of clause (viii), by striking the period at the end of clause (ix) and inserting a comma, and by inserting after clause (ix) the following new clauses:

> ``(x) the portion of the credit determined under section 30B which is attributable to the application of subsection (e)(3) thereof with respect to new qualified alternative fuel motor vehicles which are capable of being powered by compressed or liquefied natural gas, and
>
> ``(xi) the portion of the credit determined under section 30C which is attributable to the application of subsection (b) thereof with respect to refueling property which is used to store and or dispense compressed or liquefied natural gas.''.

(b) Personal Credits.--

(2) New qualified alternative fuel motor vehicles.—

Subsection (g) of section 30B is amended by adding at the end the following new paragraph:

``(3) Special rule relating to certain new qualified alternative fuel motor vehicles.--In the case of the portion of the credit determined under subsection (a) which is attributable to the application of subsection (e)(3) with respect to new qualified alternative fuel motor vehicles which are capable of being powered by compressed or liquefied natural gas--

``(A) paragraph (2) shall (after the application of paragraph (1)) be applied separately with respect to such portion, and

``(B) in lieu of the limitation determined under paragraph (2), such limitation shall not exceed the excess (if any) of--

``(i) the sum of the regular tax liability (as defined in section 26(b)) plus the tentative minimum tax for the taxable year, reduced by

``(ii) the sum of the credits allowable under subpart A and sections 27 and 30.''.

(2) Alternative fuel vehicle refueling properties.-- Subsection (d) of section 30C is amended by adding at the end the following new paragraph:

``(3) Special rule relating to certain alternative fuel vehicle refueling properties.--In the case of the portion of the credit determined under subsection (a) with respect to refueling property which is used to store and or dispense compressed or liquefied natural gas and which is attributable to the application of subsection (b)—

``(A) paragraph (2) shall (after the application of paragraph (1)) be applied separately with respect to such portion, and

``(B) in lieu of the limitation determined under paragraph (2), such limitation shall not exceed the excess (if any) of--

``(i) the sum of the regular tax liability (as defined in section 26(b)) plus the tentative minimum tax for the taxable year, reduced by

``(ii) the sum of the credits allowable under subpart A and sections 27, 30, and the portion of the credit determined under section 30B which is attributable to the application of subsection (e)(3) thereof.''.

(c) Credits May Be Transferred.--

(1) Vehicle credits.--Subsection (h) of section 30B is amended by adding at the end the following new paragraph:

``(11) Transferability of credit.--

``(A) In general.--Except as provided in subparagraph (B), a taxpayer who places in service any new qualified alternative fuel motor vehicle which is capable of being powered by compressed or liquefied natural gas may transfer the credit allowed under this section by reason of subsection (e) with respect to such vehicle through an assignment to the manufacturer, seller or lessee of such vehicle. Such transfer may be revoked only with the consent of the Secretary.

``(B) Denial of double benefit.--No assignment of a credit allowed under this section by reason of subsection (e) with respect to any new qualified alternative fuel motor vehicle which is capable of being powered by compressed or liquefied natural gas may be made under subparagraph (A) to a taxpayer who has claimed a credit under section 54G with respect to the financing of such vehicle.

``(C) Regulations.--The Secretary shall prescribe such regulations as necessary to ensure that any credit transferred under subparagraph (A) is claimed once and not reassigned by such other person.''.

(2) Infrastructure credit.--Subsection (e) of section 30C is amended by adding at the end the following new paragraph:

``(7) Transferability of credit.--

``(A) In general.--Except as provided in subparagraph (B), a taxpayer who places in service any qualified alternative fuel vehicle refueling property relating to compressed or liquefied natural gas may transfer the credit allowed under this section with respect to such property through an assignment to the manufacturer, seller or lessee of such property. Such transfer may be revoked only with the consent of the Secretary.

``(B) Denial of double benefit.--No assignment of a credit allowed under this section by reason of subsection (e) with respect to any qualified alternative fuel vehicle refueling property relating to compressed or liquefied natural gas may be made under subparagraph (A) to a taxpayer who has claimed a credit under section 54G with respect to the financing of such property.

142

``(C) Regulations.--The Secretary shall prescribe such regulations as necessary to ensure that any credit transferred under subparagraph (A) is claimed once and not reassigned by such other person.''.

(d) Effective Date.--The amendments made by this section shall apply with respect to property placed in service after the date of the enactment of this Act.

SEC. 104. MODIFICATION OF CREDIT FOR PURCHASE OF VEHICLES FUELED BY COMPRESSED NATURAL GAS OR LIQUEFIED NATURAL GAS.

(a) Increase in Credit.--Paragraph (2) of section 30B(e) (relating to applicable percentage) is amended to read as follows:

``(2) Applicable percentage.--For purposes of paragraph (1), the applicable percentage with respect to any new qualified alternative fuel motor vehicle is--

``(A) except as provided in subparagraphs (B) and (C)--

``(i) 50 percent, plus

``(ii) 30 percent, if such vehicle--

``(I) has received a certificate of conformity under the Clean Air Act and meets or exceeds the most stringent standard available for certification under the Clean Air Act for that make and model year vehicle (other than a zero emission standard), or

``(II) has received an order certifying the vehicle as meeting the same requirements as vehicles which may be sold or leased in California and meets or exceeds the most stringent standard available for certification under the State laws of California (enacted in accordance with a waiver granted under section 209(b) of the Clean Air Act) for that make and model year vehicle (other than a zero emission standard),

``(B) 80 percent, in the case of dedicated vehicles that are only capable of operating on compressed or liquefied natural gas, dual-fuel vehicles that are only capable of operating on a mixture of no less than 90 percent compressed or liquefied natural gas, and a bi-fuel vehicle that is capable of operating a minimum of 85 percent of its total range on compressed or liquefied natural gas, and

``(C) 50 percent, in the case of vehicles described subclause (II) or (III) of subsection (e)(4)(A)(i) and which are not otherwise described in subparagraph (B).
For purposes of the preceding sentence, in the case of any new qualified alternative fuel motor vehicle which weighs more than 14,000 pounds gross vehicle weight rating, the most stringent standard available shall be such standard available for certification on the date of the enactment of the Energy Tax Incentives Act of 2005.''.

(b) Increased Incentive for Natural Gas Vehicles.--Subsection (e) of section 30B (relating to new qualified alternative fuel motor vehicle credit) is amended by adding at the end the following new paragraph:

``(6) Credit values for natural gas vehicles.--In the case of new qualified alternative fuel motor vehicles with respect to vehicles powered by compressed or liquefied natural gas, the maximum tax credit value shall be--

``(A) $7,500 if such vehicle has a gross vehicle weight rating of not more than 8,500 pounds,

``(B) $16,000 if such vehicle has a gross vehicle weight rating of more than 8,500 pounds but not more than 14,000 pounds,

``(C) $40,000 if such vehicle has a gross vehicle weight rating of more than 14,000 pounds but not more than 26,000 pounds, and

``(D) $64,000 if such vehicle has a gross vehicle weight rating of more than 26,000 pounds.".

(c) Effective Date.--The amendment made by this section shall apply to property placed in service after the date of the enactment of this Act.

SEC. 105. MODIFICATION OF DEFINITION OF NEW QUALIFIED ALTERNATIVE FUEL MOTOR VEHICLE.

(a) In General.--Clause (i) of section 30B(e)(4)(A) (relating to definition of new qualified alternative fuel motor vehicle) is amended to read as follows:

``(i) which--

``(I) is a dedicated vehicle that is only capable of operating on an alternative fuel,

``(II) is a bi-fuel vehicle that is capable of operating on compressed or liquefied natural gas and gasoline or diesel fuel, or

``(III) is a duel-fuel vehicle that is capable of operating on a mixture of compressed or liquefied natural gas and gasoline or diesel fuel.".

(b) Conversions and Repowers.--Paragraph (4) of section 30B(e) is amended by adding at the end the following new subparagraph:

``(C) Conversions and repowers.--

``(i) In general.--The term `new qualified alternative fuel motor vehicle' includes the conversion or repower of a new or used vehicle so that it is capable of operating on an alternative fuel as it was not previously capable of operating on an alternative fuel.

``(ii) Treatment as new.--A vehicle which has been converted to operate on an alternative fuel shall be treated as new on the date of such conversion for purposes of this section.

``(iii) Rule of construction.--In the case of a used vehicle which is converted or repowered, nothing in this section shall be construed to require that the motor vehicle be acquired in the year the credit is claimed under this section with respect to such vehicle.".

(c) Effective Date.--The amendments made by this section shall apply to property placed in service after the date of the enactment of this Act.

SEC. 106. PROVIDING FOR THE TREATMENT OF PROPERTY PURCHASED BY INDIAN TRIBAL GOVERNMENTS.

(a) In General.--Paragraph (6) of section 30B(h) and paragraph (2) of section 30C(e) are both amended by inserting ``, or an Indian Tribal Government'' after ``section 50(b)''.

(b) Effective Date.--The amendments made by this section shall apply to property placed in service after the date of the enactment of this Act.

TITLE II--PROMOTE PRODUCTION OF NGVS BY ORIGINAL EQUIPMENT MANUFACTURERS

SEC. 201. CREDIT FOR PRODUCING VEHICLES FUELED BY NATURAL GAS OR LIQUIFIED NATURAL GAS.

(a) In General.--Subpart D of part IV of subchapter A of chapter 1 (relating to business-related credits) is amended by inserting after section 45R the following new section:

``SEC. 45S. PRODUCTION OF VEHICLES FUELED BY NATURAL GAS OR LIQUIFIED NATURAL GAS.

``(a) In General.--For purposes of section 38, in the case of a taxpayer who is an original manufacturer of natural gas vehicles, the natural gas vehicle credit determined under this section for any taxable year with respect to each eligible natural gas vehicle produced by the taxpayer during such year is an amount equal to the lesser of--

``(1) 10 percent of the manufacturer's basis in such vehicle, or

``(2) $4,000.

``(b) Aggregate Credit Allowed.--The aggregate amount of credit allowed under subsection (a) with respect to a taxpayer for any taxable year shall not exceed $200,000,000 reduced by the amount of the credit allowed under subsection (a) to the taxpayer (or any predecessor) for all prior taxable years.

``(c) Definitions.--For the purposes of this section--

``(1) Eligible natural gas vehicle.--The term `eligible natural gas vehicle' means a motor vehicle (as defined in section 30B(h)(1)) that is capable of operating on natural gas and is described in 30B(e)(4)(A).

``(2) Manufacturer.--The term `manufacturer' has the meaning given such term in regulations prescribed by the Administrator of the Environmental Protection Agency for purposes of title II of the Clean Air Act (42 U.S.C. 7521 et seq.).

``(d) Special Rules.--For purposes of this section--

``(1) In general.--Rules similar to the rules of subsections (c), (d), and (e) of section 52 shall apply.

``(2) Controlled groups.--

``(A) In general.--All persons treated as a single employer under subsection (a) or (b) of section 52 or subsection (m) or (o) of section 414 shall be treated as a single producer.

``(B) Inclusion of foreign corporations.--For purposes of subparagraph (A), in applying subsections (a) and (b) of section 52 to this section, section 1563 shall be applied without regard to subsection (b)(2)(C) thereof.

``(C) Verification.--No amount shall be allowed as a credit under subsection (a) with respect to which the taxpayer has not submitted such information or certification as the Secretary, in consultation with the Secretary of Energy, determines necessary.

``(e) Termination.--This section shall not apply to any vehicle produced after December 31, 2016.''.

(b) Credit To Be Part of Business Credit.--Section 38(b) is amended by striking ``plus'' at the end of paragraph (35), by striking the period at the end of paragraph (36) and inserting ``, plus'', and by adding at the end the following:

``(37) the natural gas vehicle credit determined under section 45R(a).''.

(c) Conforming Amendment.--The table of sections for subpart D of part IV of subchapter A of chapter 1 is amended by inserting after the item relating to section 45R the following new item:

``Sec. 45S. Production of vehicles fueled by natural gas or liquefied natural gas.''.

(d) Effective Date.--The amendments made by this section shall apply to vehicles produced after December 31, 2011.

SEC. 202. ADDITIONAL VEHICLES QUALIFYING FOR THE ADVANCED TECHNOLOGY VEHICLES MANUFACTURING INCENTIVE PROGRAM.

(a) In General.--Notwithstanding any other provision of law, a covered vehicle (as defined in subsection (b)) shall be considered an advanced technology vehicle for purposes of the advanced technology vehicle incentive program established under section 136 of the Energy Independence and Security Act of 2007 (42 U.S.C. 17013), and manufacturers and component suppliers of such covered vehicles shall be eligible for an award under such section.

(b) Definitions.--As used in this section--

(1) the term ``covered vehicle'' means a light-duty vehicle or a medium-duty or heavy-duty truck or bus that is only capable of operating on compressed or liquefied natural gas, a bi-fueled motor vehicle that is capable of achieving a minimum of 85 percent of its total range with compressed or liquefied natural gas, or a dual-fuel vehicle that operates on a mixture of natural gas and gasoline or diesel fuel but is not capable of operating on a mixture of less than 75 percent natural gas;

(2) the term ``bi-fuel vehicle'' means a vehicle that is capable of operating on compressed or liquefied natural gas and gasoline or diesel fuel; and

(3) the term ``dual-fuel vehicle'' means a vehicle that is capable of operating on a mixture of compressed or liquefied natural gas and gasoline or diesel fuel.

TITLE III--INCENTIVIZE THE INSTALLATION OF NATURAL GAS FUEL PUMPS

SEC. 301. EXTENSION AND MODIFICATION OF ALTERNATIVE FUEL VEHICLE REFUELING PROPERTY CREDIT.

(a) In General.--Subsection (g) of section 30C is amended by striking ``and'' at the end of paragraph (1), by redesignating paragraph (2) as paragraph (3), and by inserting after paragraph (1) the following new paragraph:

    ``(2) in the case of property relating to compressed or liquefied natural gas, after December 31, 2016, and".

  (b) Effective Date.--The amendments made by subsection (a) shall apply to property placed in service after the date of the enactment of this Act.

SEC. 302. INCREASE IN CREDIT FOR CERTAIN ALTERNATIVE FUEL VEHICLE REFUELING PROPERTIES.

  (a) In General.--Subsection (b) of section 30C is amended to read as follows:

  ``(b) Limitation.--The credit allowed under subsection (a) with respect to all qualified alternative fuel vehicle refueling property placed in service by the taxpayer during the taxable year at a location shall not exceed--

    ``(1) except as provided in paragraph (2), $30,000 in the case of a property of a character subject to an allowance for depreciation,

    ``(2) in the case of compressed natural gas property and liquefied natural gas property which is of a character subject to an allowance for depreciation, the lesser of--

      ``(A) 50 percent of such cost, or

      ``(B) $100,000, and

    ``(3) $2,000 in any other case.".

  (b) Effective Date.--The amendment made by this section shall apply to property placed in service in taxable years beginning after December 31, 2011.

TITLE IV--NATURAL GAS VEHICLES

SEC. 401. GRANTS FOR NATURAL GAS VEHICLES RESEARCH AND
DEVELOPMENT.

(a) Research, Development and Demonstration Programs.--The
Secretary shall provide funding to improve the performance and
efficiency and integration of natural gas powered motor vehicles and
heavy-duty on-road vehicles as part of any programs funded pursuant to
section 911 of the Energy Policy Act of 2005 (42 U.S.C. 16191) and also
with respect to funding for heavy-duty engines pursuant to section 754
of the Energy Policy Act of 2005 (42 U.S.C. 16102).

(b) In General.--The Secretary of Energy may make grants to
original equipment manufacturers of light-duty and heavy-duty natural
gas vehicles for the development of engines that reduce emissions,
improve performance and efficiency, and lower cost.

SEC. 402. SENSE OF THE CONGRESS REGARDING EPA CERTIFICATION
OF NGV RETROFIT KITS.

It is the sense of the Congress that the Environmental Protection
Agency should streamline the process for certification of natural gas
vehicle retrofit kits to promote energy security while still fulfilling
the mission of the Clean Air Act.

SEC. 403. SENSE OF THE CONGRESS REGARDING EPA AND NHTSA
REGULATION OF MEDIUM- AND HEAVY-DUTY ENGINES AND
VEHICLES.

It is the sense of the Congress that the Environmental Protection Agency new fuel economy and greenhouse gas emission regulations for medium- and heavy-duty engines and vehicles should provide incentives to encourage and reward manufacturers who produce natural gas powered vehicles. Such regulations should take into account the petroleum reductions provided by such vehicles and also quantify all greenhouse gas emission reductions provided by natural gas powered engines and vehicles.

SEC. 404. AMENDMENT TO SECTION 508 OF THE ENERGY POLICY ACT OF 1992.

(a) Repower or Converted Alternative Fueled Vehicles Defined.-- Subsection (a) of section 508 of the Energy Policy Act of 1992 (42 U.S.C. 13258) is amended by adding at the end the following new paragraph:

    ``(6) Repowered or converted.--The term `repowered or converted' means modified with a certified engine or aftermarket system so that the vehicle is capable of operating on an alternative fuel.''.

(b) Allocation of Credits.--Subsection (b) of section 508 of the Energy Policy Act of 1992 (42 U.S.C. 13258) is amended by adding at the end the following new paragraph:

    ``(3) Repowered or converted vehicles.--Not later than January 1, 2012, the Secretary shall allocate credits to fleets that repower or convert an existing vehicle so that it is capable of operating on an alternative fuel. In the case of any medium-duty or heavy-duty vehicle that is repowered or converted, the Secretary shall allocate additional credits for such vehicles if the Secretary determines that such vehicles displace more petroleum than light-duty alternative fueled vehicles. The Secretary shall include a requirement that such vehicles remain in the fleet for a period of no less than 2 years in order to continue to qualify for credit. The Secretary also shall extend the flexibility afforded in this section to Federal fleets subject to the purchase provisions contained in section 303 of this Act.".

This is a copy of S1863 of 2011. The other bill is HR 1380 of 2011 which starts on Page 133.

[Congressional Bills 112th Congress]
[From the U.S. Government Printing Office]
[S. 1863 Introduced in Senate (IS)]

112th CONGRESS

1st Session

S. 1863

To amend the Internal Revenue Code of 1986 to encourage alternative energy investments and job creation.

IN THE SENATE OF THE UNITED STATES

November 15, 2011

Mr. Menendez (for himself, Mr. Reid, Mr. Burr, and Mr. Chambliss) introduced the following bill; which was read twice and referred to the Committee on Finance

A BILL

To amend the Internal Revenue Code of 1986 to encourage alternative
energy investments and job creation.

Be it enacted by the Senate and House of Representatives of the
United States of America in Congress assembled,

SECTION 1. SHORT TITLE, ETC.

(a) Short Title.--This Act may be cited as the ``New Alternative
Transportation to Give Americans Solutions Act of 2011''.

(b) Amendment of 1986 Code.--Except as otherwise expressly
provided, whenever in this Act an amendment or repeal is expressed in
terms of an amendment to, or repeal of, a section or other provision,
the reference shall be considered to be made to a section or other
provision of the Internal Revenue Code of 1986.

(c) Table of Contents.--The table of contents for this Act is as
follows:

Sec. 1. Short title, etc.

TITLE I--PROMOTE THE PURCHASE AND USE OF NGVS WITH AN
EMPHASIS ON
    HEAVY-DUTY VEHICLES AND FLEET VEHICLES

Sec. 101. Extension and modification of new qualified alternative fuel
    motor vehicle credit.

Sec. 102. Allowance of vehicle and infrastructure credits against
    regular and minimum tax and transferability
    of credits.
Sec. 103. Modification of credit for purchase of vehicles fueled by
    compressed natural gas or liquefied natural
    gas.
Sec. 104. Modification of definition of new qualified alternative fuel
    motor vehicle.
Sec. 105. Providing for the treatment of property purchased by Indian
    tribal governments.

TITLE II--PROMOTE PRODUCTION OF NGVS BY ORIGINAL EQUIPMENT MANUFACTURERS

Sec. 201. Credit for producing vehicles fueled by natural gas or
    liquified natural gas.
Sec. 202. Additional vehicles qualifying for the advanced technology
    vehicles manufacturing incentive program.

TITLE III--INCENTIVIZE THE INSTALLATION OF NATURAL GAS FUEL PUMPS

Sec. 301. Extension and modification of alternative fuel vehicle
    refueling property credit.
Sec. 302. Increase in credit for certain alternative fuel vehicle
    refueling properties.

TITLE IV--NATURAL GAS VEHICLES

Sec. 401. Grants for natural gas vehicles research and development.

Sec. 102. Allowance of vehicle and infrastructure credits against
    regular and minimum tax and transferability
    of credits.
Sec. 103. Modification of credit for purchase of vehicles fueled by
    compressed natural gas or liquefied natural
    gas.
Sec. 104. Modification of definition of new qualified alternative fuel
    motor vehicle.
Sec. 105. Providing for the treatment of property purchased by Indian
    tribal governments.

TITLE II--PROMOTE PRODUCTION OF NGVS BY ORIGINAL EQUIPMENT MANUFACTURERS

Sec. 201. Credit for producing vehicles fueled by natural gas or
    liquified natural gas.
Sec. 202. Additional vehicles qualifying for the advanced technology
    vehicles manufacturing incentive program.

TITLE III--INCENTIVIZE THE INSTALLATION OF NATURAL GAS FUEL PUMPS

Sec. 301. Extension and modification of alternative fuel vehicle
    refueling property credit.
Sec. 302. Increase in credit for certain alternative fuel vehicle
    refueling properties.

TITLE IV--NATURAL GAS VEHICLES

Sec. 401. Grants for natural gas vehicles research and development.

(a) Business Credits.--Subparagraph (B) of section 38(c)(4) is amended by striking ``and'' at the end of clause (viii), by striking the period at the end of clause (ix) and inserting a comma, and by inserting after clause (ix) the following new clauses:

``(x) the portion of the credit determined under section 30B which is attributable to the application of subsection (e)(3) thereof with respect to new qualified alternative fuel motor vehicles which are capable of being powered by compressed or liquefied natural gas, and

``(xi) the portion of the credit determined under section 30C which is attributable to the application of subsection (b) thereof with respect to refueling property which is used to store and or dispense compressed or liquefied natural gas.''.

(b) Personal Credits.--

(1) New qualified alternative fuel motor vehicles.-- Subsection (g) of section 30B is amended by adding at the end the following new paragraph:

``(3) Special rule relating to certain new qualified alternative fuel motor vehicles.--In the case of the portion of the credit determined under subsection (a) which is attributable to the application of subsection (e)(3) with respect to new qualified alternative fuel motor vehicles which are capable of being powered by compressed or liquefied natural gas--

``(A) paragraph (2) shall (after the application of paragraph (1)) be applied separately with respect to such portion, and

``(B) in lieu of the limitation determined under paragraph (2), such limitation shall not exceed the excess (if any) of--

``(i) the sum of the regular tax liability (as defined in section 26(b)) plus the tentative minimum tax for the taxable year, reduced by

``(ii) the sum of the credits allowable under subpart A and sections 27 and 30.''.

(2) Alternative fuel vehicle refueling properties.-- Subsection (d) of section 30C is amended by adding at the end the following new paragraph:

``(3) Special rule relating to certain alternative fuel vehicle refueling properties.--In the case of the portion of the credit determined under subsection (a) with respect to refueling property which is used to store and or dispense compressed or liquefied natural gas and which is attributable to the application of subsection (b)--

``(A) paragraph (2) shall (after the application of paragraph (1)) be applied separately with respect to such portion, and

``(B) in lieu of the limitation determined under paragraph (2), such limitation shall not exceed the excess (if any) of--

``(i) the sum of the regular tax liability (as defined in section 26(b)) plus the tentative minimum tax for the taxable year, reduced by

``(ii) the sum of the credits allowable under subpart A and sections 27, 30, and the portion of the credit determined under section 30B which is attributable to the application of subsection (e)(3) thereof.''.

(c) Credits May Be Transferred.--

(1) Vehicle credits.--Subsection (h) of section 30B is amended by adding at the end the following new paragraph:

''(11) Transferability of credit.--

''(A) In general.--Except as provided in subparagraph (B), a taxpayer who places in service any new qualified alternative fuel motor vehicle which is capable of being powered by compressed or liquefied natural gas may transfer the credit allowed under this section by reason of subsection (e) with respect to such vehicle through an assignment to the manufacturer, seller or lessee of such vehicle. Such transfer may be revoked only with the consent of the Secretary.

''(B) Regulations.--The Secretary shall prescribe such regulations as necessary to ensure that any credit transferred under subparagraph (A) is claimed once and not reassigned by such other person.''.

(2) Infrastructure credit.--Subsection (e) of section 30C is amended by adding at the end the following new paragraph:

''(7) Transferability of credit.--

''(A) In general.--Except as provided in subparagraph (B), a taxpayer who places in service any qualified alternative fuel vehicle refueling property relating to compressed or liquefied natural gas may transfer the credit allowed under this section with respect to such property through an assignment to the manufacturer, seller or lessee of such property. Such transfer may be revoked only with the consent of the Secretary.

``(B) Regulations.--The Secretary shall prescribe such regulations as necessary to ensure that any credit transferred under subparagraph (A) is claimed once and not reassigned by such other person.''.

(d) Effective Date.--The amendments made by this section shall apply with respect to property placed in service after the date of the enactment of this Act.

SEC. 103. MODIFICATION OF CREDIT FOR PURCHASE OF VEHICLES FUELED BY COMPRESSED NATURAL GAS OR LIQUEFIED NATURAL GAS.

(a) Increase in Credit.--Paragraph (2) of section 30B(e) is amended to read as follows:

``(2) Applicable percentage.--For purposes of paragraph (1), the applicable percentage with respect to any new qualified alternative fuel motor vehicle is--

``(A) except as provided in subparagraphs (B) and (C)--

``(i) 50 percent, plus

``(ii) 30 percent, if such vehicle--

``(I) has received a certificate of conformity under the Clean Air Act and meets or exceeds the most stringent standard available for certification under the Clean Air Act for that make and model year vehicle (other than a zero emission standard), or

``(II) has received an order
certifying the vehicle as meeting the
same requirements as vehicles which may
be sold or leased in California and
meets or exceeds the most stringent
standard available for certification
under the State laws of California
(enacted in accordance with a waiver
granted under section 209(b) of the
Clean Air Act) for that make and model
year vehicle (other than a zero
emission standard),

``(B) 80 percent, in the case of dedicated vehicles
that are only capable of operating on compressed or
liquefied natural gas, dual-fuel vehicles that are only
capable of operating on a mixture of no less than 90
percent compressed or liquefied natural gas, and a bi-
fuel vehicle that is capable of operating a minimum of
85 percent of its total range on compressed or
liquefied natural gas, and

``(C) 50 percent, in the case of vehicles described
subclause (II) or (III) of subsection (e)(4)(A)(i) and
which are not otherwise described in subparagraph (B).
For purposes of the preceding sentence, in the case of any new
qualified alternative fuel motor vehicle which weighs more than
14,000 pounds gross vehicle weight rating, the most stringent
standard available shall be such standard available for
certification on the date of the enactment of the Energy Tax
Incentives Act of 2005.''.

(b) Increased Incentive for Natural Gas Vehicles.--Subsection (e) of section 30B is amended by adding at the end the following new paragraph:

``(6) Credit values for natural gas vehicles.--In the case of new qualified alternative fuel motor vehicles with respect to vehicles powered by compressed or liquefied natural gas, the maximum tax credit value shall be--

``(A) $7,500 if such vehicle has a gross vehicle weight rating of not more than 8,500 pounds,

``(B) $16,000 if such vehicle has a gross vehicle weight rating of more than 8,500 pounds but not more than 14,000 pounds,

``(C) $40,000 if such vehicle has a gross vehicle weight rating of more than 14,000 pounds but not more than 26,000 pounds, and

``(D) $64,000 if such vehicle has a gross vehicle weight rating of more than 26,000 pounds.''.

(c) Effective Date.--The amendment made by this section shall apply to property placed in service after the date of the enactment of this Act.

SEC. 104. MODIFICATION OF DEFINITION OF NEW QUALIFIED ALTERNATIVE FUEL MOTOR VEHICLE.

(a) In General.--Clause (i) of section 30B(e)(4)(A) is amended to read as follows:

``(i) which--

``(I) is a dedicated vehicle that is only capable of operating on an alternative fuel,

``(II) is a bi-fuel vehicle that is

capable of operating on compressed or
liquefied natural gas and gasoline or
diesel fuel, or

``(III) is a duel-fuel vehicle that
is capable of operating on a mixture of
compressed or liquefied natural gas and
gasoline or diesel fuel.''.

(b) Conversions and Repowers.--Paragraph (4) of section 30B(e) is
amended by adding at the end the following new subparagraph:

``(C) Conversions and repowers.--

``(i) In general.--The term `new qualified
alternative fuel motor vehicle' includes the
conversion or repower of a new or used vehicle
so that it is capable of operating on an
alternative fuel as it was not previously
capable of operating on an alternative fuel.

``(ii) Treatment as new.--A vehicle which
has been converted to operate on an alternative
fuel shall be treated as new on the date of
such conversion for purposes of this section.

``(iii) Rule of construction.--In the case
of a used vehicle which is converted or
repowered, nothing in this section shall be
construed to require that the motor vehicle be
acquired in the year the credit is claimed
under this section with respect to such
vehicle.''.

(c) Effective Date.--The amendments made by this section shall
apply to property placed in service after the date of the enactment of
this Act.

SEC. 105. PROVIDING FOR THE TREATMENT OF PROPERTY PURCHASED BY INDIAN
TRIBAL GOVERNMENTS.

(a) In General.--Paragraph (6) of section 30B(h) and paragraph (2) of section 30C(e) are both amended by inserting ``, or an Indian Tribal Government'' after ``section 50(b)''.

(b) Effective Date.--The amendments made by this section shall apply to property placed in service after the date of the enactment of this Act.

TITLE II--PROMOTE PRODUCTION OF NGVS BY ORIGINAL EQUIPMENT MANUFACTURERS

SEC. 201. CREDIT FOR PRODUCING VEHICLES FUELED BY NATURAL GAS OR LIQUIFIED NATURAL GAS.

(a) In General.--Subpart D of part IV of subchapter A of chapter 1 is amended by inserting after section 45R the following new section:

``SEC. 45S. PRODUCTION OF VEHICLES FUELED BY NATURAL GAS OR LIQUIFIED NATURAL GAS.

``(a) In General.--For purposes of section 38, in the case of a taxpayer who is an original manufacturer of natural gas vehicles, the natural gas vehicle credit determined under this section for any taxable year with respect to each eligible natural gas vehicle produced by the taxpayer during such year is an amount equal to the lesser of--

``(1) 10 percent of the manufacturer's basis in such vehicle, or

``(2) $4,000.

``(b) Aggregate Credit Allowed.--The aggregate amount of credit allowed under subsection (a) with respect to a taxpayer for any taxable year shall not exceed $200,000,000 reduced by the amount of the credit allowed under subsection (a) to the taxpayer (or any predecessor) for all prior taxable years.

``(c) Definitions.--For the purposes of this section--

``(1) Eligible natural gas vehicle.--The term `eligible natural gas vehicle' means a motor vehicle (as defined in section 30B(h)(1)) that is capable of operating on natural gas and is described in 30B(e)(4)(A).

``(2) Manufacturer.--The term `manufacturer' has the meaning given such term in regulations prescribed by the Administrator of the Environmental Protection Agency for purposes of title II of the Clean Air Act (42 U.S.C. 7521 et seq.).

``(d) Special Rules.--For purposes of this section--

``(1) In general.--Rules similar to the rules of subsections (c), (d), and (e) of section 52 shall apply.

``(2) Controlled groups.--

``(A) In general.--All persons treated as a single employer under subsection (a) or (b) of section 52 or subsection (m) or (o) of section 414 shall be treated as a single producer.

``(B) Inclusion of foreign corporations.--For purposes of subparagraph (A), in applying subsections (a) and (b) of section 52 to this section, section 1563 shall be applied without regard to subsection (b)(2)(C) thereof.

``(C) Verification.--No amount shall be allowed as

a credit under subsection (a) with respect to which the
taxpayer has not submitted such information or
certification as the Secretary, in consultation with
the Secretary of Energy, determines necessary.

``(e) Termination.--This section shall not apply to any vehicle
produced after December 31, 2016.''.

(b) Credit To Be Part of Business Credit.--Section 38(b) is amended
by striking ``plus'' at the end of paragraph (35), by striking the
period at the end of paragraph (36) and inserting ``, plus'', and by
adding at the end the following:

``(37) the natural gas vehicle credit determined under
section 45S(a).''.

(c) Conforming Amendment.--The table of sections for subpart D of
part IV of subchapter A of chapter 1 is amended by inserting after the
item relating to section 45R the following new item:

``Sec. 45S. Production of vehicles fueled by natural gas or liquefied
natural gas.''.

(d) Effective Date.--The amendments made by this section shall
apply to vehicles produced after December 31, 2011.

SEC. 202. ADDITIONAL VEHICLES QUALIFYING FOR THE ADVANCED
TECHNOLOGY VEHICLES MANUFACTURING INCENTIVE PROGRAM.

(a) In General.--Notwithstanding any other provision of law, a
covered vehicle (as defined in subsection (b)) shall be considered an
advanced technology vehicle for purposes of the advanced technology
vehicle incentive program established under section 136 of the Energy
Independence and Security Act of 2007 (42 U.S.C. 17013), and

manufacturers and component suppliers of such covered vehicles shall be eligible for an award under such section.

(b) Definitions.--As used in this section--

(1) the term ``covered vehicle'' means a light-duty vehicle or a medium-duty or heavy-duty truck or bus that is only capable of operating on compressed or liquefied natural gas, a bi-fueled motor vehicle that is capable of achieving a minimum of 85 percent of its total range with compressed or liquefied natural gas, or a dual-fuel vehicle that operates on a mixture of natural gas and gasoline or diesel fuel but is not capable of operating on a mixture of less than 75 percent natural gas;

(2) the term ``bi-fuel vehicle'' means a vehicle that is capable of operating on compressed or liquefied natural gas and gasoline or diesel fuel; and

(3) the term ``dual-fuel vehicle'' means a vehicle that is capable of operating on a mixture of compressed or liquefied natural gas and gasoline or diesel fuel.

TITLE III--INCENTIVIZE THE INSTALLATION OF NATURAL GAS FUEL PUMPS

SEC. 301. EXTENSION AND MODIFICATION OF ALTERNATIVE FUEL VEHICLE REFUELING PROPERTY CREDIT.

(a) In General.--Subsection (g) of section 30C is amended by striking ``and'' at the end of paragraph (1), by redesignating paragraph (2) as paragraph (3), and by inserting after paragraph (1) the following new paragraph:

``(2) in the case of property relating to compressed or liquefied natural gas, after December 31, 2016, and''.

(b) Effective Date.--The amendments made by subsection (a) shall apply to property placed in service after the date of the enactment of this Act.

SEC. 302. INCREASE IN CREDIT FOR CERTAIN ALTERNATIVE FUEL VEHICLE REFUELING PROPERTIES.

(a) In General.--Subsection (b) of section 30C is amended to read as follows:

``(b) Limitation.--The credit allowed under subsection (a) with respect to all qualified alternative fuel vehicle refueling property placed in service by the taxpayer during the taxable year at a location shall not exceed--

    ``(1) except as provided in paragraph (2), $30,000 in the case of a property of a character subject to an allowance for depreciation,

    ``(2) in the case of compressed natural gas property and liquefied natural gas property which is of a character subject to an allowance for depreciation, the lesser of--

        ``(A) 50 percent of such cost, or

        ``(B) $100,000, and

    ``(3) $2,000 in any other case.''.

(b) Effective Date.--The amendment made by this section shall apply to property placed in service in taxable years beginning after December 31, 2011.

TITLE IV--NATURAL GAS VEHICLES

SEC. 401. GRANTS FOR NATURAL GAS VEHICLES RESEARCH AND DEVELOPMENT.

(a) Research, Development and Demonstration Programs.--The Secretary shall provide funding to improve the performance and efficiency and integration of natural gas powered motor vehicles and heavy-duty on-road vehicles as part of any programs funded pursuant to section 911 of the Energy Policy Act of 2005 (42 U.S.C. 16191) and also with respect to funding for heavy-duty engines pursuant to section 754 of the Energy Policy Act of 2005 (42 U.S.C. 16102).

(b) In General.--The Secretary of Energy may make grants to original equipment manufacturers of light-duty and heavy-duty natural gas vehicles for the development of engines that reduce emissions, improve performance and efficiency, and lower cost.

SEC. 402. SENSE OF THE CONGRESS REGARDING EPA CERTIFICATION OF NGV RETROFIT KITS.

It is the sense of the Congress that the Environmental Protection Agency should further streamline the process for certification of natural gas vehicle retrofit kits to promote energy security while still fulfilling the mission of the Clean Air Act.

SEC. 403. AMENDMENT TO SECTION 508 OF THE ENERGY POLICY ACT OF 1992.

(a) Repower or Converted Alternative Fueled Vehicles Defined.-- Subsection (a) of section 508 of the Energy Policy Act of 1992 (42 U.S.C. 13258) is amended by adding at the end the following new paragraph:

``(6) Repowered or converted.--The term `repowered or converted' means modified with a certified or approved engine or aftermarket system so that the vehicle is capable of operating on an alternative fuel.".

(b) Allocation of Credits.--Subsection (b) of section 508 of the Energy Policy Act of 1992 (42 U.S.C. 13258) is amended by adding at the end the following new paragraph:

> ``(3) Repowered or converted vehicles.--Not later than January 1, 2012, the Secretary shall allocate credits to fleets or covered persons that repower or convert an existing vehicle so that it is capable of operating on an alternative fuel. In the case of any medium-duty or heavy-duty vehicle that is repowered or converted, the Secretary shall allocate additional credits for such vehicles if the Secretary determines that such vehicles displace more petroleum than light-duty alternative fueled vehicles. The Secretary shall include a requirement that such vehicles remain in the fleet for a period of no less than 2 years in order to continue to qualify for credit. The Secretary also shall extend the flexibility afforded in this section to Federal fleets subject to the purchase provisions contained in section 303 of this Act.".

TITLE V--TRANSIT SYSTEMS

SEC. 501. FEDERAL SHARE OF COSTS FOR EQUIPMENT FOR COMPLIANCE WITH CLEAN AIR ACT.

Section 5323(i) of title 49, United States Code, is amended--
> (1) in paragraph (1)--
>> (A) in the paragraph heading, by striking ``and clean air act'';
>> (B) in the first sentence, by striking ``or vehicle-related'' and all that follows through ``Clean

Air Act"; and

(C) by striking ``those Acts'' each place it
appears and inserting ``the Americans with Disabilities
Act of 1990 (42 U.S.C. 12101 et seq.)'';

(2) by redesignating paragraph (2) as paragraph (3); and

(3) by inserting after paragraph (1) the following:

``(2) Equipment for compliance with clean air act.--

``(A) In general.--A grant for a project to be
assisted under this chapter that involves acquiring
vehicle-related equipment or facilities (including
clean fuel or alternative fuel vehicle-related
equipment or facilities) for purposes of complying with
or maintaining compliance with the Clean Air Act (42
U.S.C. 7401 et seq.) shall be made for--

``(i) 100 percent of the net project cost
of the equipment or facilities attributable to
compliance with that Act for any amounts of not
more than $75,000; and

``(ii) 90 percent of the net project cost
of the equipment or facilities attributable to
compliance with that Act for any amounts of
more than $75,000.

``(B) Costs.--The Secretary shall have discretion
to determine, through practicable administrative
procedures, the costs of equipment or facilities
attributable to compliance with the Clean Air Act (42
U.S.C. 7401 et seq.).''.

SEC. 502. NATURAL GAS TRANSIT INFRASTRUCTURE INVESTMENT.

(a) Establishment.--The Secretary of Transportation shall establish and administer a program to encourage the development of natural gas fueling infrastructure to be used by transit agencies.

(b) Use.--Funding provided under the program may be used for the purpose of building new or expanded fueling facilities, if the expansion is for the purposes of fueling additional buses with natural gas.

(c) Competitive Grants.--The Secretary shall--

(1) administer the funding providing under the program on a competitive basis; and

(2) award funding after an evaluation of project proposals that includes--

(A) the overall quantity of petroleum to be displaced over the life of the proposed project;

(B) the amount of private funding or local funding that is available to offset the cost of the project; and

(C) the technical and economical feasibility of the project.

(d) Authorization of Appropriations.--There is authorized to be appropriated to carry out this section $100,000,000, to remain available until expended.

TITLE VI--USER FEES

SEC. 601. USER FEES.

(a) Liquefied Natural Gas.--Clause (ii) of section 4041(a)(2)(B) is amended by striking ``24.3 cents per gallon'' and inserting ``the sum of the Highway Trust Fund financing rate and the Natural Gas

Transportation Incentives financing rate".

(b) Compressed Natural Gas.--The second sentence of subparagraph (A) of section 4041(a)(3) is amended by striking ``18.3 cents per energy equivalent of a gallon of gasoline'' and inserting ``the sum of the Highway Trust Fund financing rate and the Natural Gas Transportation Incentives financing rate".

(c) Highway Trust Fund Financing Rate and Natural Gas Transportation Incentives Financing Rate.--Subsection (a) of section 4041 is amended by adding at the end the following new paragraph:

``(4) Highway trust fund financing rate and natural gas transportation incentives financing rate.--For purposes of this title--

``(A) Highway trust fund financing rate.--The term `Highway Trust Fund financing rate' means--

``(i) with respect to liquefied natural gas, 24.3 cents per gallon, and

``(ii) with respect to compressed natural gas, 18.3 cents per energy equivalent of a gallon of gasoline.

``(B) Natural gas transportation incentives financing rate.--

``(i) In general.--The term `Natural Gas Transportation Incentives financing rate' means--

``(I) with respect to liquefied natural gas, the applicable amount per gallon, and

``(II) with respect to compressed natural gas, the applicable amount per

energy equivalent of a gallon of
gasoline.

``(ii) Applicable amount.--For purposes of
clause (i), the applicable amount shall be
determined in accordance with the following
table:

``Calendar year	Applicable amount
2014...	2.5 cents
2015...	2.5 cents
2016...	5 cents
2017...	5 cents
2018...	10 cents
2019...	10 cents
2020...	12.5 cents
2021...	12.5 cents
2022 and thereafter......................................	zero.

``(iii) Exemption for fuel dispensed from
certain property.--In the case of liquefied
natural gas or compressed natural gas dispensed
from property for which a credit under section
30C(b)(3) would be allowable, the applicable
amount for any calendar year is zero.''.

  (c) Natural Gas Transportation Incentives Financing Rate Deposited in General Fund.--Paragraph (4) of section 9503 is amended by striking ``or'' at the end of subparagraph (C), by striking the period at the end of subparagraph (D)(iii) and inserting ``or'', and by adding at the end the following new subparagraph:

    ``(E) section 4041 to the extent attributable to the Natural Gas Transportation Incentives financing rate.''.

     <all>

APPENDIX F
WESTERN RESEARCH
(Since 1987)

Western Research Institute (WRI) is a non-profit Idaho corporation serving the general public, private clients, the press and government. The main purpose for which WRI was founded, was and still is, investigations and reporting. The Institute is first of all a provider of facts. Most of what the Institute does is centered on this mission. Facts compiled are issued in the form of reports or books.

In addition to compiling and publishing reports and books, WRI is involved in the collection and dissemination of news and opinion with special emphasis on the law, the courts and the legal profession. WRI is an independent non-partisan reporter of facts. It has no agenda except freedom of information and holding public officials and others accountable for what they say for public consumption.

WRI has compiled and published a number of reports and books on various subjects including law, government, technology, philosophy and communication.

Its main mission remains serving the public and the press with information that would otherwise not see the light of day. Only a tiny fraction of the events in this country and around the world are ever reported on. Very little activity of what transpires in the myriad of government agencies in this country ever reaches the public eye.

WRI was incorporated in January of 1987 and has been in continuous existence since then. This can be verified by clicking on the business entities section of the Idaho Secretary of State. Further information is at the WRI web site at wrii.org

APPENDIX G
ABOUT THE AUTHORS

JERRY FENNING

Jerry Fenning has been working on a private basis for almost a quarter of a century in researching and developing hypotheses regarding peace and war. His initial manuscripts were dated back to 1986 and concentrated on describing the concept of artigraph as it applied to public measurements of changes in the global nuclear stockpile over a 100 year period of time.

He continued to expand his explanations and theories of peace and war until it reached its current state of accomplishment. The subject matter was very narrow in scope at the beginning phase of development but the project has evolved into a much more comprehensive sweep of how culture affects peace and war activities. The goal is to provide a significant contribution as to the consequences of how people can live their lives on an individual and collective basis. Much more evidence and discovery remains to be ascertained in order to verify the theories but publication of the document should encourage other researchers to offer their contributions. Mr. Fenning believes that the breadth of the manuscript is sufficient for publication and has a good chance to achieve its stated purpose.

Although one of Mr. Fenning's life long goals involved research and ultimately publication of matters relating to peace and war, his professional career of about 36 years primarily consisted of work as a rehabilitation counselor who provided vocational services to adults with disabilities in the public sector. He obtained two master's degrees, one in psychology and the other as a California public school counselor (pupil personnel services), which enabled him to occupy this position. He assumed leadership positions in various professional and community organizations throughout the years. These included activities in the National Rehabilitation Association in Southern California, membership in the board of trustees of a local synagogue and participation in social action endeavors. He has more time to devote to his private research since his retirement from full time work as a rehabilitation counselor.

Mr. Fenning has invested in companies espousing goals in alternative energy. Westport Innovative Inc. and Gasfrac Inc. are two such investments that he has had either past and/or present financial positions in.

His Web site and blog is **http://cfpeaceandwar.com/**

He has an additional blog on the companion web site at **http://natgas-rpt.com**

CHARLES HOPPINS

Charles Hoppins is a semi-retired journalist with a passion for philosophy enhanced by formal graduate study and a lot of research. He has worked for The Associated Press and a number of newspapers as writer, city editor, state editor and wire editor.

Reporters working for news organizations are under constant pressure to produce copy and meet deadlines. The stories they cover often have serious ramifications that are dealt with only on a superficial basis. Troubled by the lack of time to do serious in depth reporting on a number of stories, he conceived of the idea of a research institution that would have the funding and ability to cover stories that would otherwise not see the light of day.

In 1987, he was a principle in the founding of Western Research Institute (WRI). From its inception WRI set out to do in-depth reporting on a number of issues. It has done reports and published books on technology, law, government, philosophy and communication.

For the past number of years he has been involved in the study of the prevalence of certain philosophical positions and how those positions affect society, human behavior and public policy. He has completed a number of books on the results, the last involving precepts of dialogue and general principles of communication.

The outcome of these studies, is the publication of his latest book, *Rules for Public Discourse*, on how to hold lawmakers, candidates for public office and others accountable for what they say for public consumption.   The book is currently available on Amazon and Barnes and Noble.

As part of WRI's commitment to public interest research, he has in collaboration with Jerry Fenning embarked on an in-depth research project on *How America Can Stop Importing Foreign Oil.*

-30-

www.ingramcontent.com/pod-product-compliance
Lightning Source LLC
Chambersburg PA
CBHW022056210326
41519CB00054B/534